Mit besten grüßen
und wünschen in Deiner
Neuen heimat. Alles liebe
Mechtilde

Mit den besten Wünschen
für das neue „Abenteuer"
Kirsten Liebe

... Liebe und Gute
... Gabi, John, Katharina
+ Alexandra Weiand

Spirit in the Grass

The Cariboo Chilcotin's Forgotten Landscape

Die allerbesten Wünsche
und alles Gute in
Eurer neuen Heimat.
Evi & Gebhard

Alles Gute weiterhin
und bleibt vor allen
Dingen gesund!
Evi + Tim

Alles Gute und
vergiß uns nicht!
Herzlichst Deine Siegrid

Mit guten Wünschen
alles Gute, Deine Rosée

Best Wishes,
Chris Harris

Liebe Lydia!
Wir wünschen Euch in
Nanaimo noch viele gesunde und glückliche Jahre
Margit + Robert

Alles liebe u. Gruß
Renate

in Erinnerung an
schöne Horseflyzeiten -
alles Liebe - vor allem Gesundheit
für Euch
Deine Urda

Alles Gute
von Erwin + Hildegard

Viele gute Wünsche
Elke
Hans Seibert

Chris Harris

with Ordell Steen, Kristi Iverson & Harold Rhenisch

RASSLANDS CONSERVATION COUNCIL OF BRITISH COLUMBIA

Spirit in the Grass

The Cariboo Chilcotin's Forgotten Landscape

Chris Harris

with Ordell Steen, Kristi Iverson & Harold Rhenisch

GRASSLANDS CONSERVATION COUNCIL OF BRITISH COLUMBIA

Country Light Publishing
105 Mile Ranch, BC

CANADIAN CATALOGUING IN PUBLICATION DATA

Edited by Harold Rhenisch
Designed by Bill Horne
Printed and bound in China

Maps by Bill Horne; base maps courtesy of the Grasslands Conservation Council of B.C.

LIBRARY AND ARCHIVES CANADA CATALOGUING IN PUBLICATION
Harris, Chris, 1939-
Spirit in the Grass: the Cariboo Chilcotin's forgotten landscape / Chris Harris; with Ordell Steen, Kristi Iverson & Harold Rhenisch.
(Discover British Columbia Books)
ISBN 978-0-9685216-8-7 (bound). --ISBN 978-0-9685216-9-4 (pbk.)
1. Grasslands--British Columbia--Cariboo Region--Pictorial works.
2. Grasslands--British Columbia--Cariboo Region. 3. Biotic communities--British Columbia--Cariboo Region--Pictorial works. 4. Biotic communities--British Columbia--Cariboo Region. I. Steen, O. A. II. Iverson, Kristi
III. Title. IV. Series.
FC3845.C3H373 2007 577.409711'75
C2007-903464-0

All images in this book have been made in-camera. Care has been taken to match this book's printed colours as closely as possible to the original film or digital capture.

PUBLISHED BY
Country Light Publishing

To order limited edition prints of any images in this book, or to order this book and others in the series *Discover British Columbia Books*™, contact:

Country Light Publishing
Box 333, 108 Mile Ranch, B.C.
Canada V0K 2Z0
T 250-791-6631, T 1-800-946-6622, F 250-791-6671
www.chrisharris.com
photography@chrisharris.com

Your comments are welcome.
The sale of this book will assist in the conservation of British Columbia's grasslands.

Previous page: Middle Grasslands in Junction Sheep Range Park.

Right: Every footstep through the bluebunch wheatgrass is a conversation.

Dedication

For much of her life, Anna Roberts has explored and documented the grasslands of the Cariboo-Chilcotin, amassing and sharing an otherwise largely unimagined body of knowledge. For the mantle of care, understanding and inspiration she has thrown over the grasslands, this book is dedicated to her.

In the grasslands, one discovers oneself in the earth — and in the sky.

Acknowledgements

This grasslands book project has received so much support and good will from so many people that I hardly know where to begin to express my sincere gratitude. If I have forgotten anyone, it is a fault of words, not of my heart, and I humbly apologize.

In 2003, a vision was born to create a book that would combine scientific knowledge and photographic imagery so unexpected that it would ignite imaginations and inspire a groundswell of public support for the conservation of the Cariboo-Chilcotin grasslands. This book is the fulfillment of that vision. I humbly thank my partners in this enterprise, the ecologists Ordell Steen and Kristi Iverson, who gave me the detailed technical knowledge that led me into the marvels of the grass and all the creatures that live from it and support it, as well as my friend, Mike Duffy, who supported the original vision and shared his inspiration. Deep thanks go as well to the poet, essayist and editor Harold Rhenisch, for the book's poems and cultural history, as well as for guiding and inspiring us with his patience and vision as we purified the book together from a virtual mountain of research, notes and images.

I also deeply thank my partner in life, Rita Giesbrecht, for sharing in my inspiration and even leading it, and for supporting me in my passion for capturing the grass. I thank my designer, Bill Horne, for his consummate skill at bringing all the threads of this project into the form of a living book, and the First Nations people, who have given us all the grass and in so many senses the stories and terms capable of speaking of it.

In addition to the work and guidance of my partners, the vision for this book has been nurtured by an amazingly diverse group of generous individuals. At times when I was really wondering if I could continue, both friends and people unknown to me contributed in so many different ways, each making a very important contribution to the book. Some contributed technical expertise or sponsored equipment (including a long telephoto lens for bird photography). Many more answered our appeal for cash to cover expenses, purchased photographic prints and imagery, passed on my newsletter for others to read, and joined us on our adventures in the grasslands. I would especially like to thank my artist sister Jane O'Malley who underwrote a huge tally of expenses to allow us to continue.

The confidence of the people named here has carried us through the many challenges that four years of work on this project have presented. They have our deep thanks. If this book accomplishes the aims of the original vision, and this sacred and precious landscape is preserved intact, that achievement will be theirs. We name them here in acknowledgement and in appreciation of their support and their vision. Major contributors include:

Anita & Mani Bader of Ten-ee-ah Lodge, Jean Berry, Barry Bolton, Anthony Cecil and Don Savjord of Bridge Creek Estate Ltd., Mike Duffy & Leslie Duffy, Michael & Chris Heimdallson of ComputerSmith, Dean Hull, Donna M. Iverson Personal Law Corporation, Kristi Iverson and Ken Mackenzie, Phil Jones, Dr. Philip A. Jones, Nicole Kipfer, Claire Kujundzic and Bill Horne, Carol MacKenzie, John & Evva McCarvill, Shayne Middleton, Jane O'Malley, Jai Mukerji, Jeffrey Newman of JN Webdesign, Susie Parrent, Anna Roberts, Tom and Mary Smith, Olli Sovio, Ordell Steen, Dieter Stenschke, Benjamin Stephen, Olaf Strumper, Dave Doubek of Tall Oak Forestry, Pat Teti and the Williams Lake Field Naturalists Society.

This project was completed through the generous support of The Nature Conservancy Canada and the Ministry of Environment.

Ministry of Environment

Further friends and supporters of the grasslands include: Angela George, Dorothy Johnson, Teyah Dan and Herman Paul of Esketemc, Mary Aulin, Karen Barger, Sage Birchwater, Mike & Darlene Calyniuk, Cheryl & Merlin Chatwin, Mark Collette, Chris Czajkowski, Glen Davidson, Bruno Delesalle, Sean Donahue, Teresa Donk, Frank Dwyer, Tiffany Edwardsen, Jen, Tom and Sarah Ellison, Janet & Ron Evans, Kathy Fell, Lillian Fletcher, Donna Ford, Etelka Gillespie, Jill Gould, Jackie Haverty, Bob Haywood-Farmer, Peter & Ursula Helfer, Bill Henwood, Grant Hoffman of Riske Creek Ranch, Marilyn Hogg, Joyce, John, Ellen, Hattie and Sam Holmes of Empire Valley Ranch, Jean Marc Hourier, Rick Howie, Jeanne Hull, Jordon & Rhiana Hull, Kris Hull, Kim Jaatinen, Brenda and Jack Jenkins, Marvin Kempston, Robert Glen Ketchum, Leslie Grace Lewington, Dennis Lloyd, Joan Loeken, Karen Longwell of the *Williams Lake Tribune*, Graham MacGregor, Alan McLean, Alison McLean, Sheril Mathews, Doug and Marie Mervin of Alkali Lake Ranch, Howard Miller, Norman Newberry, Maddi Newman, Jean Oke, Susie Parrent, Roger Packham, Brian & Lily Payton, Ted Peterson, Virginia Pettman, Larry Ramstead of the Gang Ranch, George Reifel, Diane, Anassa and Leandra Rhenisch, Gina Roberts, Mike Sarell, Sherry Stewart, Shelly and John Somerville, Jacky & Dennis Trelenberg, Jordon Tucker, Frances Vyse, Gail Wallin, Cliff Ware, Anna Wärje and Richard Williams, Karen Wiebe, Anita Willis of *British Columbia Magazine*, Sandy Winston, Karen & Sam Wrinkle, and other friends, no less generous, who wish to remain anonymous.

All of these people have taken this project on as their own. It is not a journey one man or woman could have made alone. This is our journey. It is a journey into the spirit of the grass.

Table of Contents

For centuries, water draining towards the Fraser River has left bold and intricate patterns in the Lower Grasslands.

The Grasslands of British Columbia

Grasslands are landscapes where the vegetation is dominated by native grasses and herbaceous flowering plants, often in association with dryland shrubs. Wetlands and small forests add habitat diversity.

British Columbia's grasslands are part of the temperate grassland biome, which once covered about 8% of the world's land area. Today it is the most altered biome on the planet, largely replaced by extensive cultivation, invasive plants, urban development, and other human disturbances.

British Columbia contains the largest remaining area of North America's Intermountain Grassland ecosystem, which once extended northward from eastern Washington, Oregon and western Idaho. Vegetation of this ecosystem is dominated by bunchgrasses, especially bluebunch wheatgrass, often in association with abundant big sagebrush.

Grasslands are the most threatened major ecosystem of British Columbia, even more threatened than old-growth coastal rainforests. They cover less than 1% of the province, yet support almost one-third of its known threatened and vulnerable plant and animal species. Expanding human populations, attracted by warm valleys, scenic landscapes, and agricultural potentials, continue to whittle away at remaining grasslands.

Within British Columbia, the southern intermountain grasslands, especially those in the Okanagan Valley, are the most studied and most imperilled. Much less well known are the intermountain grasslands of the Cariboo-Chilcotin region, which occur at the ecosystem's northern limits. Only 5% of these grasslands have been lost to cultivation and urban development, invasive plants are relatively scarce, and wildlife roam over entire natural landscapes.

Our goal in this book is to increase awareness of the incredible beauty and ecological richness of the Cariboo-Chilcotin grasslands. It is a landscape that deserves greater attention, not only because it represents much of what has been lost elsewhere, but also because it is a unique grassland in its own right and deserving of exploration, study, and protection.

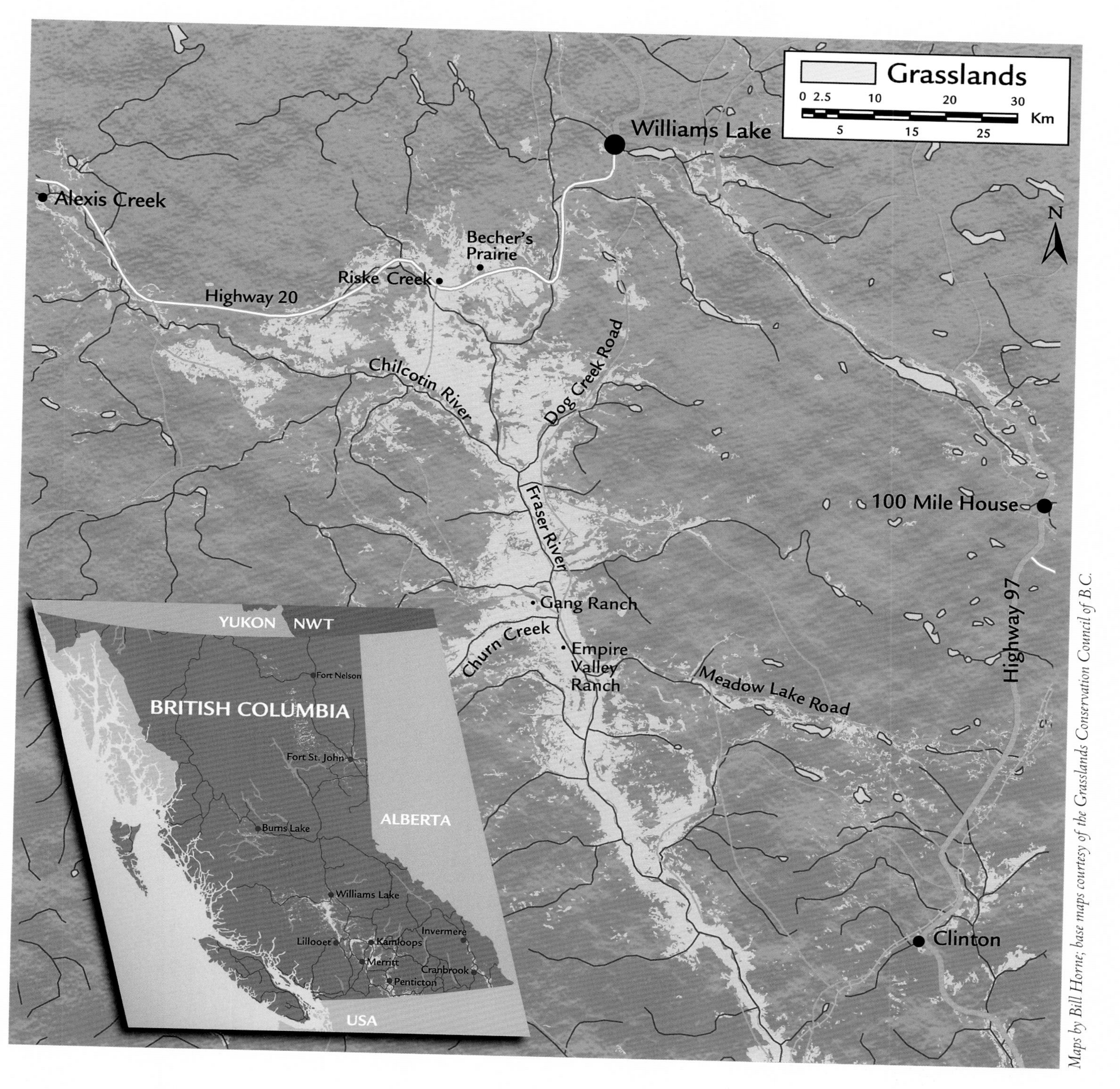

Maps by Bill Horne; base maps courtesy of the Grasslands Conservation Council of B.C.

Inset: Grasslands of British Columbia
Above: Grasslands of the Cariboo-Chilcotin

Foreword

As Chair of the Grasslands Conservation Council of British Columbia (GCC), I am genuinely pleased to introduce a truly exceptional book. *Spirit in the Grass — The Cariboo-Chilcotin's Forgotten Landscape* represents an exquisite and personal statement of passion and poetry. Chris Harris's elegant images will undoubtedly spur viewers to appreciate and hopefully to cherish one of British Columbia's most extraordinary grassland landscapes. Moreover, the Natural History section, authored by GCC ecologists, provides an informative and clearly-written scientific overview of the Cariboo-Chilcotin grasslands. A reader moving through the book becomes increasingly aware that culture emerges from the landscape — that history flows from the landscape and that people and society are inextricably part of it. For the professional and amateur photographer, Chris Harris offers insights into photography and how he met the challenge to make his "images of the rolling hills and subtle pastel colours ones that would be impossible to forget."

Spirit in the Grass predictably provides an objective statement regarding the threats facing British Columbia's incredible grasslands. It also offers a solution: "To save our grasslands, we must all change the way we locate and design housing developments, find means to keep working ranches working, map and treat invasive plants, and remove encroaching trees from this dwindling ecosystem. These are practical and reasonable steps, and they are within our means."

Part of the mission of the Grasslands Conservation Council is to foster a greater understanding of and appreciation for the ecological, social, economic, and cultural importance of British Columbia grasslands. Our support of *Spirit in the Grass* represents a vital step in a long-term campaign by the GCC to promote public awareness of the values and need for conservation of British Columbia grasslands. Chris Harris writes that he hopes "this book will serve as a source of inspiration." I believe that he and the Grasslands Conservation Council have succeeded admirably.

— Michael Pitt, Chair
Grasslands Conservation Council
of British Columbia

Below the grassland tabletops of Churn Creek Protected Area lie deep ravines where California bighorn sheep find water and reprieve from heat and light.

Hikers entering the Lower Grasslands.

A true sign of renewed life in the grasslands is the song of a western meadowlark talking to its mate through the spring air.

The Spirit of the Grass

At the heart of the Cariboo-Chilcotin, there is grass. The wind is blowing there as I write this, lightly lifting the curls of the wheatgrass, perhaps shivering the yellow petals of the salsify, and stirring the dust around the roots of the big sagebrush. As you read this, the wind is building dunes and moving rivers of heat. It never stops.

In three years devoted almost exclusively to photographing the grasslands that spill up the valleys of the Fraser and Chilcotin rivers, I've learned that these lands are always changing. Every moment here is new. The grassland that lay over the Gang Ranch five minutes ago, or over the Chilcotin River Junction, where this morning, yesterday, or four years ago California bighorn sheep scrambled up and down the clay hoodoos with unparalleled agility, are writing themselves right now into the body of the earth. The golden grass of Becher's Prairie, and the Gang Ranch benches that catch the light today like vast solar panels, and caught it a thousand years ago — and ten thousand years ago — are only the leading moment of an endless record of light, water, stone and air, as they are shaped by a complex of glacial-era landforms into a tapestry of interlocking microclimates. Some of these microclimates are no larger than the width of a grass blade. Others are the size of a city lot. Yet others span square kilometres.

In some angles of the Fraser and Chilcotin river benchlands, the wind rises and falls from valley wall to valley wall like waves breaking on an ocean shore. In others, it is powered by tiny, local changes of light and shadow, in the tilt of planes of grass or clay, or in the angles of channels of rock, that reflect tiny, self-generating climates of shade, heat, snow, fire, and the flow of water through bedrock. In this extreme climate, even the snowdrifts of deep winter can be read in the pattern of flowers and grass during the following summer. For ten thousand years the plants growing here, where it scarcely seems possible to grow, have recorded such changing patterns — a record that can be read in an instant.

I've walked through the grasslands in snow and ice, thunderstorms, rain, wind, and fire. I've tramped across ancient lichens in heat that turned my skin to leather, and have camped on open benchlands to capture the dawn. I've run crazily, madly, through the grass, chasing the light, as it shifted with the speed of the earth swinging around the sun, and have witnessed the birth of the moon out of a fold in the earth. In fact, what started as a quest to find a way to photograph an unknown and endangered landscape and to put my art to a use greater than the illustration of tourism brochures, has grown into me. The sound of the meadowlark is now a part of me, and the grasslands now centre my life.

I'd like to take you on my journey. I'd like you to kneel with me here among insects, living in tiny landscapes that could be tide pools on a distant shore, and to brush with me through patterns of big sagebrush that follow old alluvial flows like mussels lining cracks across stone laid bare by a receding tide. I'd like you to step with me under Douglas-firs that have been standing alone for centuries in lightning, starlight and wind, and to pass into the hush of aspens, sprouting from a single root system into complex communities, like ships moving across the swells of vast oceans. I'd like you to scramble with me on narrow trails through canyons of stone, in such bright colours and ever-changing drama that I've thought, surely, my eyes would grow saturated with their light, and if I went out to the grasslands to photograph them one more time I'd break the spell and would begin to repeat myself. I never have. Every time I've travelled to the grasslands, they have been a new landscape. From this experience, I've learned how to live on this earth. With my photographs, I would like to share this experience with you.

I was photographing in the grass-level world, when I heard a rock tumble down a cliff across the valley. I looked up and saw a solitary sheep that had left her herd for a tiny morsel of green vegetation in the middle of an almost vertical cliff wall. For some inscrutable reason, out of all the vast miles of available food, this sheep felt drawn to this small bush on this vertical mud cliff. Her tenacity was inspiring, although the risk/reward ratio was certainly not a human one!

California bighorn sheep

The luminous oxbows of the Chilcotin River have their headwaters amidst the high, glaciated peaks of the Coast Mountains.

And this astonishing landscape is so close to home. I am constantly amazed that it lies only two hours from my house, that my home is, in fact, among its fringes, that I can step out my door, walk ten paces, and touch the grass. I am continually humbled that I never knew this until three years ago, when I took Grasslands Conservation Council director Mike Duffy on a canoe trip through the Bowron Lakes, and we got to talking about the land, and the idea for this book came to us as our canoes rode the water and clouds formed in the clear air on the leeward sides of the mountains. In all of my years guiding in the unparalleled British Columbia wilderness of mountains, lakes, glaciers, and rivers, I never saw that we had one of the ecological wonders of the world so close to home.

Now, I am deeply moved. The canyon grasslands of the Cariboo-Chilcotin belong to one of the rarest of all ecosystems: large, intact temperate grasslands. I've walked upon them and have lived within them for weeks at a time. I've seen them as few have seen them, deep in the night, before sunrise, in snow and rain, and in every season. I'm honoured to have become a part of this ecosystem, and to have been given a chance to ensure it lives on as long as the earth.

This is not just any ecosystem, either. The grasslands and savannahs of Africa, the small colonies and islands of trees at the edge of the grass, where Australopithecus and our other ancestors merged our visual and cognitive capacities with our ability for movement, and which underlie my art as a photographer, is kin to this ecosystem. This is our home. Our ancestors first appeared in the African Savannah about 2.6 million years ago. Eventually, they brought controlled fire to grasslands, and maintained them as cultural artifacts — right into our time. Because of them, the Cariboo-Chilcotin grasslands hold our living memory. If we lose them, we lose our souls.

We who live in British Columbia should be telling the world that what has appeared lost, what has receded into ancient time, has not been lost yet — not quite. Over much of the earth, it remains only in scraps, and rags, memories, and shadows, yet here it is still alive. In Australia, a continent once bountiful

Long before the glaciers, this land was created in fire. Here above Churn Creek, the rocks and soil still carry the effects of that heat.

To communicate both the power and the delicacy of the Chilcotin River landscape, I made this image on a day with diffuse light to bring out colour, and a long lens to capture the towering spirit of the silt bluffs.

with grass, ecologists now often define pristine grasslands as areas the size of a large city lot. We are indeed blessed.

When I first travelled to the grasslands, I thought, surely, that the time had come for us to once again show respect for all forms of life — human, plant, animal, and the planet herself. In all of my journeys from one end of the grasslands to the other, in every gorge, along every slope and plateau, with cactus in my shins, flicking farther up my legs with every footstep, and with grass seeds in my socks like barbs of iron, with dust in my eyes and raindrops falling in scorching heat and landing on my skin like drops of ice, I have discovered nothing that could detract from this vision. In fact, my vision has only deepened, both as a photographer and as a human being.

It has become my passion to pass on this unexpected, priceless legacy that has come to us through accidents of history, isolation, and pure good fortune. My dream is that we can join together here to rebuild our connections with the earth, deepen our sense of what it is to be human, and accept the responsibilities that go with that.

Just as these canyon grasslands are one of the last vestiges of the great seas of grass that once covered much of the planet, we have come to the edge of our journey now. This is British Columbia's last chance. There will not be another one. We have been given the unexpected gift of this one last opportunity to rediscover who we are, to remember what we have known, and to be whole, across the entire span of our time on earth. The lives of future generations are part of that span.

As I walk across the grasslands,
I walk with them.

— *Chris Harris*

Left: Two hikers are about to cross a deer trail in one of the last pristine bunchgrass grasslands of this scale in the world.

Overleaf: As the wind rushed up to meet the speeding clouds and rain swept over the grass, I started with single images, but quickly switched to a panoramic format to capture the feeling of rapid movement over the trees.

This was my first photograph for the project, based on simple lines and tonal contrast to take advantage of the subtlety of the flat light. I felt like a photographic print was developing before my eyes. The Grasslands Awareness Book Project was under way. I was thrilled.

Spirit in the Winter

Journal Entry: February 10, 2004. Evening.

This has been Day One of the Grasslands Awareness Book Project. I headed out this morning to the grasslands west of Williams Lake, excited but apprehensive. My vision for the book was as yet unclear and I had never before photographed the grasslands in winter. As I drove along the icy road, my eyes searched the land, but I saw nothing. The day was dull and the light was flat.

Where and how would I begin? I had images in my mind but I could see that they didn't exist. I soon began to realize that I had to let go of all my preconceptions. I had to get myself into 'the present,' to see what was actually out there. Only then did the landscape slowly begin to unfold.

— *Chris Harris*

As the intensity of light began to increase, smaller worlds began to come into definition. I became aware of animal tracks and became fascinated with their journeys and the stories they told.

As I drove this tiny road on another pilgrimage to the Lower Grasslands near Churn Creek, a dusting of snow revealed a complexity of shadows, and snow that seemed to bring the light right into the earth itself.

Above: As I passed through a north-facing band of shadow and frost, I was charmed by light reflecting off stalks of grass clothed in ice crystals.

Below: Searching deeper within the ice-coated grass, I suddenly entered a new and unlimited universe — a galaxy of stars, throwing their light into the Milky Way. At first, I thought it would be rare to catch such moments. Over time, I learned that they were the rule, and gave myself to them.

I snowshoed to the top of the Farwell Canyon sand dune, where a Douglas-fir snag had been overwhelmed by sand long ago. To capture its prominence, I tried various perspectives, until two warmly lit rabbitbrush shrubs in the foreground lead my eyes to the tree, balanced between earth and sky.

The story of why a coyote followed the ridgeline of a frozen, snow-covered dune, where there would certainly be nothing to hunt, still eludes me.

Sunset at Cougar Point, overlooking
the Mid-Fraser River Canyon.

Photographer's Vision

My vision for the Cariboo-Chilcotin grasslands was to create a visually-driven, content-rich coffee table book that would celebrate the beauty of the grasslands biome through the art of photography. My goal was to promote greater public awareness of grasslands, that would lead to their ultimate conservation.

When I began my visual journeys into the grasslands, I wanted to see freshly and deeply. I consciously decided to ignore the names of all the grasses, sagebrush, lichens, and bugs around me. Without familiar labels and categories, I trusted that every subject would lead me to art. I wanted to make images of the rolling hills and subtle pastel colours that would be impossible to forget.

In my experience as a photographer, the feelings in the grasslands have been uniquely difficult to capture. Without the bold shapes and colours of glaciers, mountains, oceans, and forests, I put my self into the landscape instead.

To make these images, I poured intense energy into arranging the elements of visual design — both the natural design of my subjects and the designs I created by careful placement. Discovering a fresh and exciting perspective and finding the magic qualities of light that would show my subject as if for the very first time often meant waiting hours, or returning to my subject time and time again.

When photography is only a means of documenting subjects around us, results are often images void of spirit. Such images serve useful purposes, but not the purpose of this project. To complete this book, I set out to capture and communicate both the spirit of the grasslands and of myself, and to pass them on through my imagery. To a photographer, this transfer of energy and vision is the ultimate reward.

— *Chris Harris*

Red outcroppings were sacred to First Nations peoples, and many village sites appear beneath one. With a patterned crust like the hide of an ancient animal, this one looks alive and watchful.

Sagebrush colonizing rock avalanche chutes in the eastern valley wall of the Fraser River.

Humans have an instinctive need to enter the land. Here, the descent of water through Churn Creek Canyon draws Mike Duffy and I deep into a journey — both into the heart of the grasslands and into ourselves.

Churn Creek Adventure

Each stage of descent into the landscape is a stage in a passage into our souls. It is as if somewhere, at the mouth of the creek, perhaps, we will have reached the goal of our journey.

People and water share a profound capacity for movement. Amidst giant boulders that have reached the limit of their journey, Mike Duffy begins to cross the cold water of the creek.

Once you've walked through the land you are never the same. The landscape has entered you and can't be left behind. Mike Duffy and I came here to spend the night. We were not disappointed.

After our two-day hike in Churn Creek canyon, Mike Duffy and I returned exhausted to our vehicles. Since darkness was setting in, I thought my photography for the day was over. Never for a moment did I expect that one of the most unforgettable moments of the trip was about to take place. I was just untying my boot laces when I looked up and saw a halo of light rising above the horizon, far away across the Fraser River Canyon. Suddenly I realized it was a full moon evening. Back on my feet, I ran for my camera and longest lens. I had to work quickly. At this magnification, it's amazing how rapidly the moon moves across the viewfinder. By the end, I was working by touch alone, in pitch dark. If I hadn't known my equipment so intimately, I would have missed these amazing shots. When the shoot was over, I was both exhausted and elated!

FULL MOON OVER SAGEBRUSH

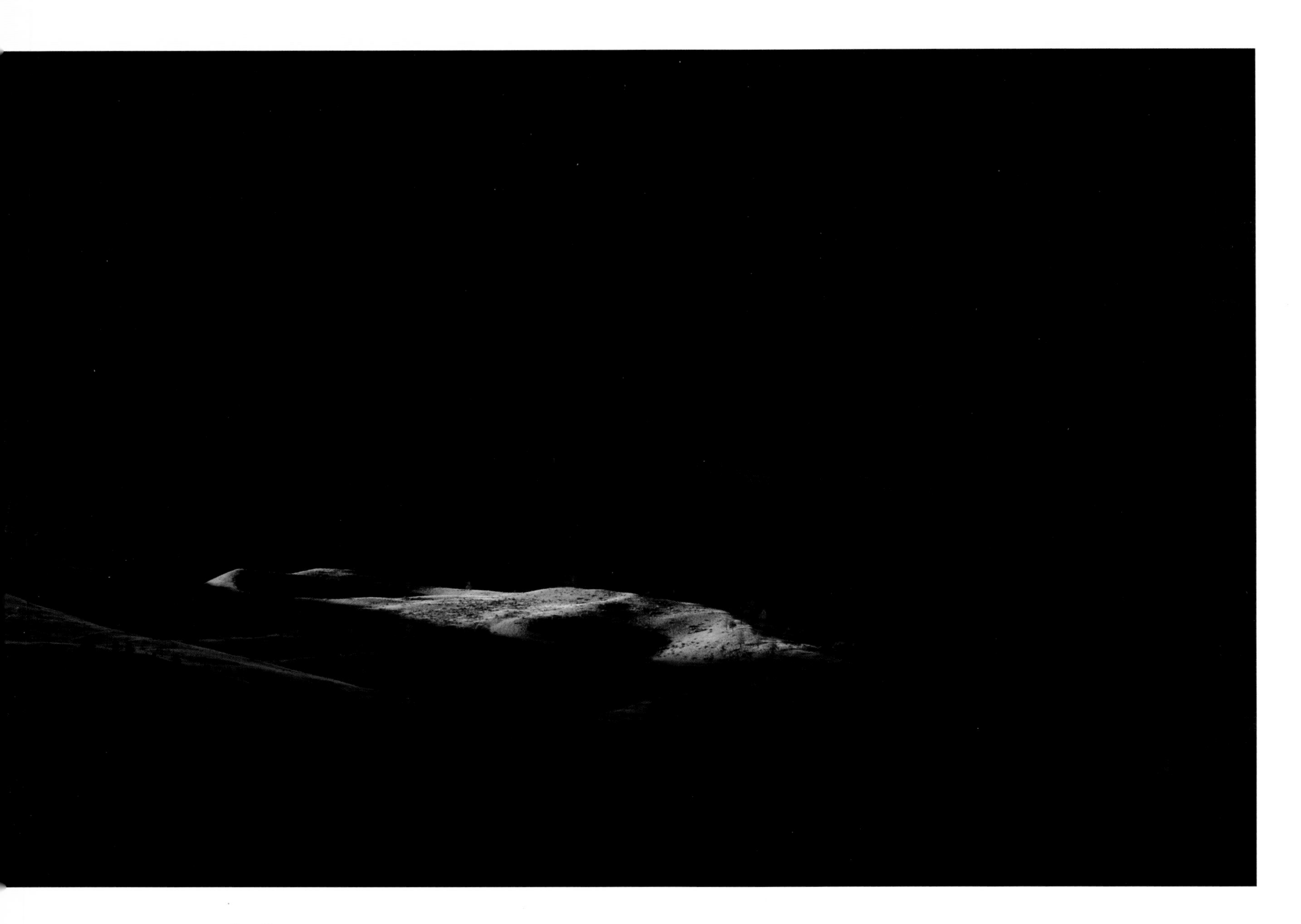

My most profound experience during this project was a series of nights spent out in the grasslands high above the Fraser River Canyon to await first light. I would walk out in the late evenings, when all the visual elemenets of the landscape slept — all except for the creator of all those elements, the river itself. And while I slept, it continued to work, carving the canyon still further.

First Light over the Grasslands

Up before dawn, I would await the magic. The first light would come quietly, as if slowly pushing the darkness away. Terrace by terrace, the sagebrush and grasses and all of their colour and texture would reappear.

Within an hour, all the hidden features of the landscape would be revealed in the warmth and energy of a new day. Each time, I felt energized for yet another day of possibility and opportunity.

The site of the great slide that temporarily blocked the Chilcotin River at Farwell Canyon in 2005, viewed in the fall of 2006.

Left: Earlier in the year I watched a red fox slowly and steadily slip through this intricate network of crevasses on its way down to the river for water, following paths that melt water had taken before it.

Below: To walk here is like walking in pure physics.

Once I discovered the ability of rain to bring drama out of monochrome landscapes, I looked forward eagerly to spring showers, and would race to Farwell Canyon for the colours and textures that they would bring out of the silt cliffs.

Abstracting the Landscape

Alone for days on end out on the grasslands, I found myself entering the spirit of trees and grass.

A group of hikers return home from a scorching day in the big sagebrush benchlands above the Fraser River.

The palette of these textured slopes awaits a new generation of painters.

A gentle slope on Becher's Prairie turns golden-yellow with the early fall colours of a porcupinegrass and spreading needlegrass community. This upper grassland community is unique on the planet.

No landscape is ever predictable or familiar. As a photographer, I pray for constant change in weather, because with change comes the possibility of dramatic light, and with dramatic light the chance to see correspondences that would otherwise be hidden.

Over time, the action of heat and cold, wet and damp, have transformed these stones to powder. Underneath them, the earth is cracking apart as well. I like to think that such little piles of colour gave the world's original artists the inspiration to express themselves by making paint and applying it to surfaces.

Red-tailed hawks, such as this juvenile at the Gang Ranch, are often seen riding updrafts through the canyonlands of the Fraser and Chilcotin rivers.

At the end of a long day, hikers slowly cross the vast deer range of Churn Flats towards the plumetting cliffland habitat of California bighorn sheep.

A few metres beyond these hikers, this fingered ridge drops down vertically more than two hundred fifty metres to tiny Churn Creek. The crumbling pillared cliff in this photograph forms the opposing canyon wall.

It was a spectacular morning. As the sun burned through the heavy mist, the scenery changed by the second. I remember moving quickly to find as many different compositions as I could before the enchantment ended. As I chased the light, I also remember silently cursing all the awkward gear I was carrying. When I returned to camp I noticed a gentleman with a camera sitting in a wheelchair. He said he had been watching me photograph and that he too was a photographer, but after a crippling accident he now had to place himself in a specific location and then wait for the light to come to him. I will never again complain about the weight of my camera gear. He was an absolute inspiration.

CHASING THE LIGHT

In the grasslands, no shot is repeatable. I walked all afternoon to shoot this view in the evening light. On my way back in the dark, I retraced my steps in the moonlight. When I returned in the late fall to catch the light in a different season, the late afternoon October sun had already set behind a mountain and this whole scene was in shadow.

Evening Light

Among my favorite subjects to photograph in the grasslands are tall snags — trees that have died of old age or have been scarred by lightning or fire.

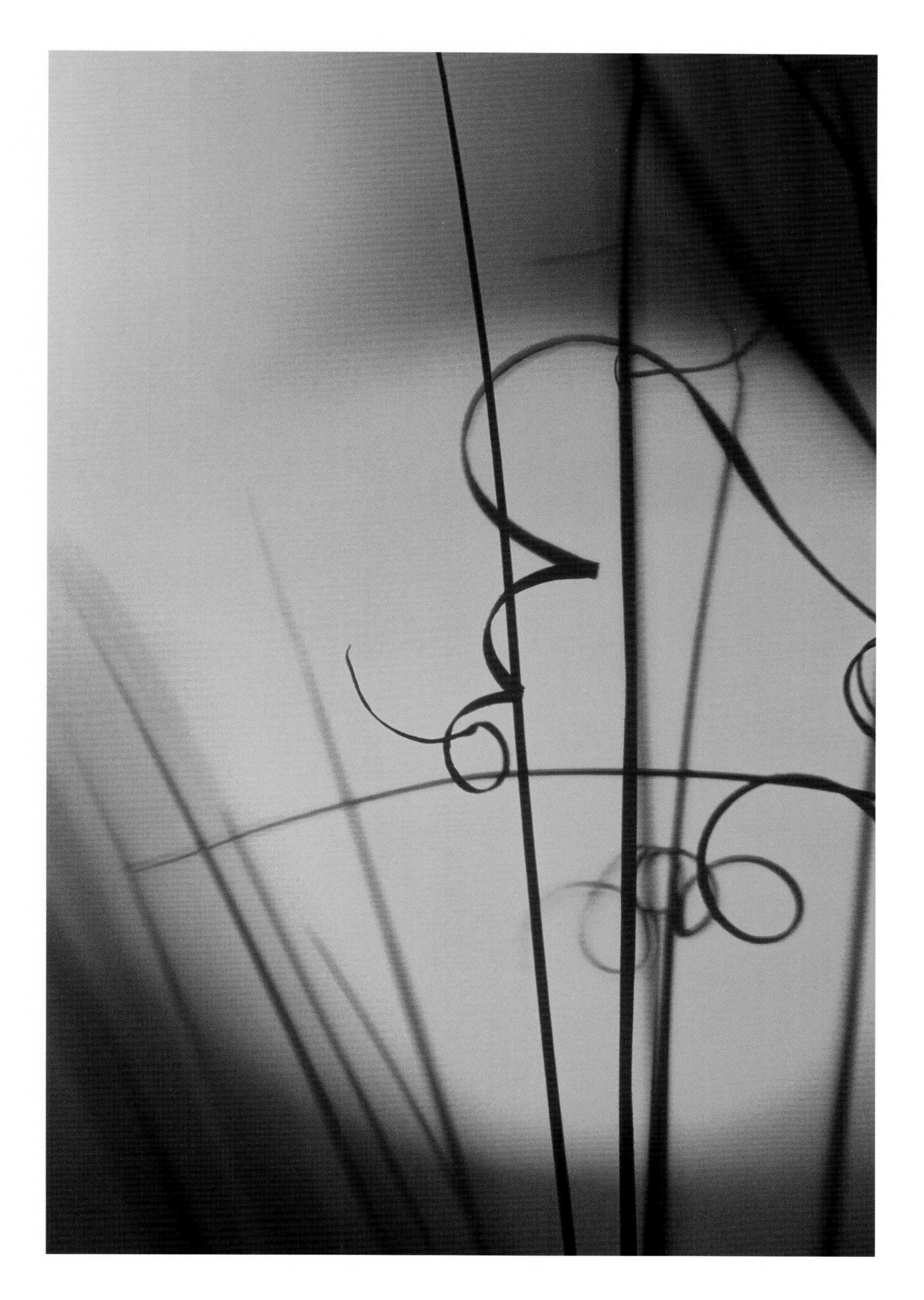

Bluebunch wheatgrass is a key indicator species of the Cariboo-Chilcotin grasslands. Its most noticeable characteristic is the curled leaves from the previous year's growth.

When I stand on these heights above the Chilcotin River, I feel how ancient life is on this earth.

The Path of the Grass

From the river that drains the sky into the sea,
as green as a leaf strung with aphids and rain,
from the gravel that the sockeye salmon
drink as they climb the ladder of darkness
into clouds of purple-mouthed lilies,
from sage root and the lung that breathes fire,
cicadas stringing necklaces of smoke
and spiders climbing the pollens of ten thousand grasses,
from all that was once broken and is now whole,
in a sky above the sky and an earth below the soil,
let me break open a salt bed and taste the sun,
let me now find the river that drains into the Pacific
and the ants that string deep swells with the snow,
for everything that was whole and is now broken
enters the land surprised, but then enters further,
as a man leaves his first footstep yet never leaves it,
as he walks towards his last but never arrives.

—*Harold Rhenisch*

Just a few moments after sunrise, the warm light does not yet have the height to completely soften the last pools of the night. The land must turn to face it.

A rolling bunchgrass blanket covers the Churn Creek Protected Area.

From time to time, storms roll quickly over the landscape, bringing moments of dramatic light and colour. I was drawn to this storm composition because of the tuft of bunchgrass in the foreground, bending lightly with the wind. The open field of blue light to the west — and the distant Pacific Ocean — indicates that this storm, too, will quickly pass.

The atmospheric mood of this image was due to the large Chilcotin Forest Fire of 2005. I made this image in memory of the Edwards family, who lost their famous homestead at Lonesome Lake to that fire.

Fire is a creator. A day after a planned burn had swept across the golden hues of old grass on Becher's Prairie, I returned to discover this compelling landscape. The entire area had been transformed to a dynamic balance of blue, white, gold and black. The tufts of smoke that make this landscape look volcanic are mounds of smoking cow dung.

RENEWAL

On one hike through the grasslands, I walked into a large area of burnt big sagebrush. I shot several rolls of film here, entranced by the visual dance of fluid greens and frozen blacks.

BURNT SAGEBRUSH

In the Lower Grassland near Farwell Canyon lies the largest sand dune in British Columbia. As the sun crosses the sky, the shadow of the dune reverses direction. The effect is as if the dune is holding the sun on a line of energy: gentle, silent, and eternal. The shape of the dune, the line of the ridge, and the texture of the rippled sand change continuously.

FARWELL CANYON SAND DUNE

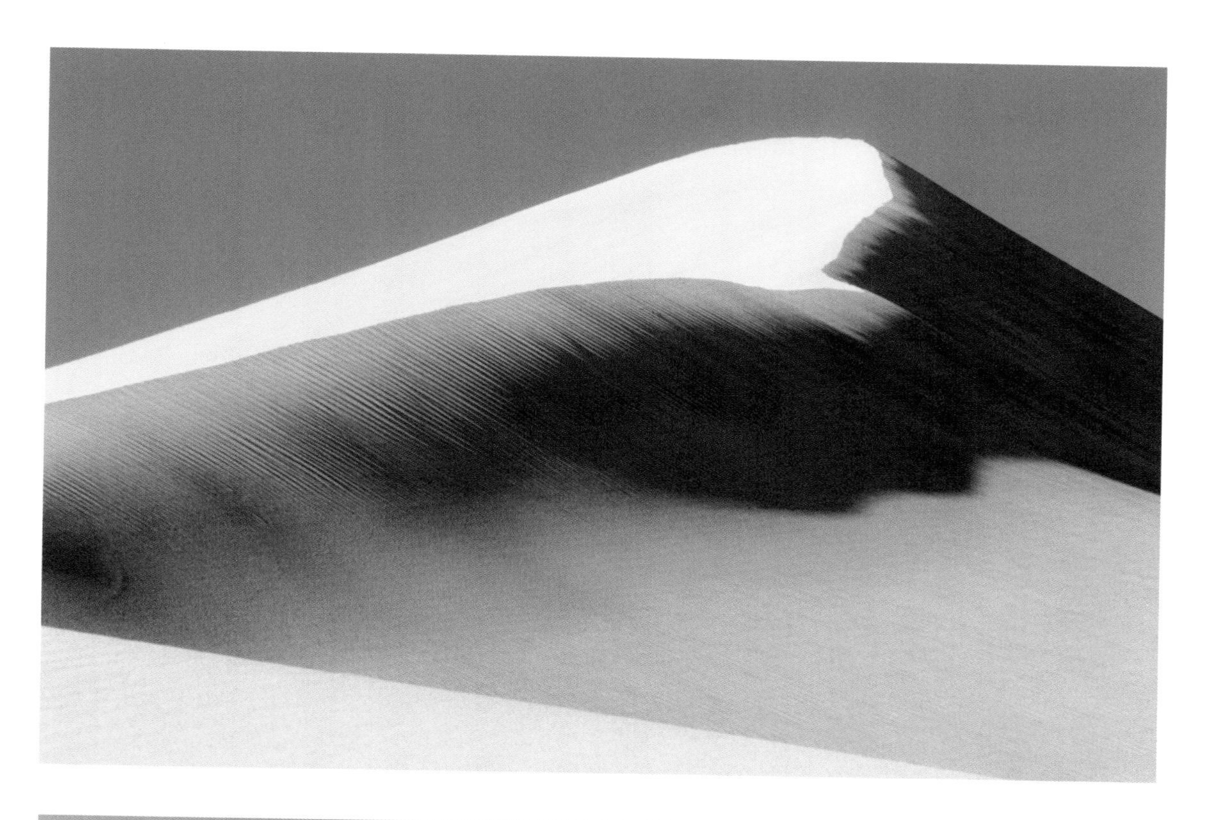

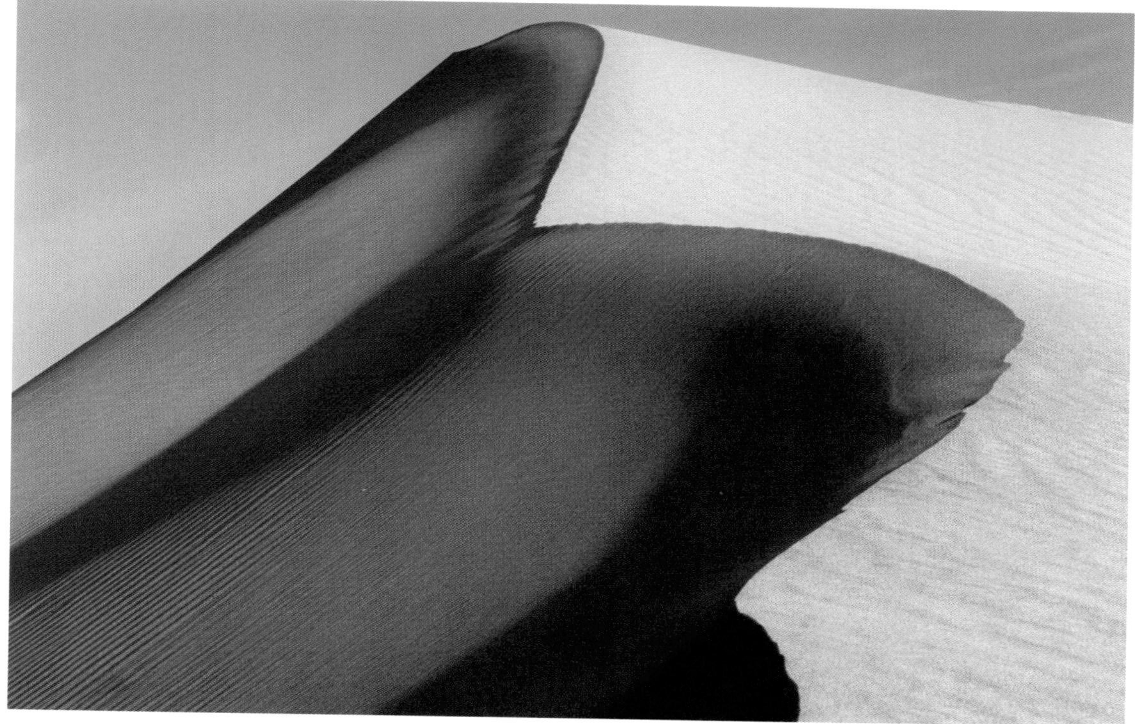

Among silt crevasses, light reveals the persistent power of the meltwater that flows off the edge of the plateau.

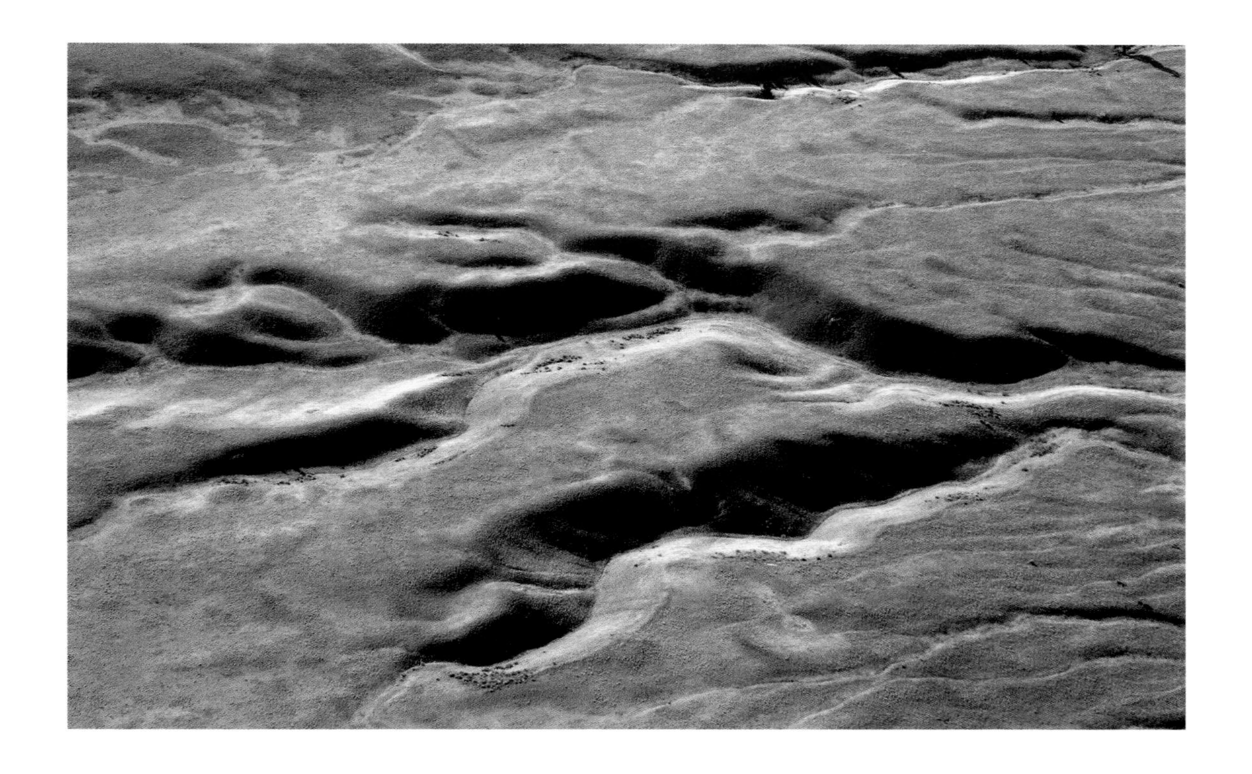

Above: To walk in the grasslands is to follow the paths of ancient water. What were once vast, deep, cold lakes and channels are now sinks and basins for heat.

Below: The Chilcotin River merges with the Fraser, bringing its load of glacial flour from the high peaks of the Coast Mountains to the Fraser River's long journey to the sea.

A series of unique pond environments stretches across the width and breadth of the grasslands. Some ponds have high levels of dissolved carbonate salts that remain on the ground like snow after the water evaporates.

Alkaline flats are prime habitats for a small number of highly adapted plants and animals. Glasswort is one of the most salt tolerant of all plants. In September, its red seeds are vibrant against the white salt crust.

Spring is an exciting time on the grasslands. In many places, meltwater is trapped by a layer of frozen ground. As soon as the layer of frost has thawed, many lakes disappear — such as this one.

Fall provides new photographic possibilities. The trees and bushes that follow riparian corridors between the broken benchlands leap out with colour against the sun-washed hues of the hills.

In the grasslands, road patterns intuitively mimic the flow of plants and water. Whenever I come across a new one, I can't resist following it.

Where water pools at the base of this small wetland, an aspen grove glows in the fall light. From mid-August to the end of September the different aspen clones turn colour at different times.

Light fills the inner world of an aspen grove in late September.

In the fall, these intricate patterns of silvery big sagebrush and rabbit-brush, bordering riparian areas of autumn-coloured aspen and a smattering of red rose, go on for hundreds of kilometres.

The sagebrush goes to seed and its branch tops droop over like flowing lace.

Overleaf: Bluebunch wheatgrass under a stormy sky.

Stepping Into the Sky

A cloud
of dust stings as I
step from my
car into yellow
salsify
blooms folding into
noon, then

passes, under a sky
so near it
vanishes. The sudden
wheatgrasses feel
like voices
speaking with their
hands. As a light

wind lifts their heavy
stamens, a yellow
blush of pollen
sings, and
rises, and there
is only
the sun

in the heads
of wheatgrass; it
burns from below my
feet and
catches
on the earth,
in its blue dust field.

— *Harold Rhenisch*

Land gave us our first beauty. When I saw this mauve and green landscape, I felt as if a planet of water and grass were opening before me at the beginning of time. Then I noticed the animal tracks cutting laterally across the image, and realized just how long this process has been going on.

Intermountain grasslands occupy the valley of the Mid-Fraser River. Lower and Middle Grasslands *(see page 119)* fill the valley, while Upper Grasslands occur on the plateau seen in the upper right. Eroded terraces adjacent to the river are formed of sediments deposited in glacial lakes and rivers that filled the valley about 10,000 years ago.

The Natural History

In the Cariboo-Chilcotin region of central British Columbia, there is a grassland that is one of the great ecological jewels of western North America. Nestled in the warm valleys of the Fraser River and its tributaries, this landscape of canyons and plateaus, covered by bunchgrasses, sagebrush and cactus has been relatively little affected by cultivation, urban development or invasive plants. A few wetlands and aspen groves dot the landscape, especially on gentle slopes. It is a landscape where the song of meadowlarks and the calls of long-billed curlews pierce the air and California bighorn sheep stand warily against the skyline as a gentle wind blows freely through the native grass.

The Cariboo-Chilcotin grassland is the northernmost extension of a vast intermountain grassland that once covered the Columbia Basin of Washington, Oregon and Idaho, and extended northward into the dry valleys of southern British Columbia. The feature that characterizes this large grassland area and distinguishes it from the grasslands of the Great Plains is bunchgrass, especially bluebunch wheatgrass. Today most of this intermountain grassland has been replaced by cultivated croplands and weed infested pastures. In the United States, less than 5% of the original native grassland is still present. Fortunately, in the Cariboo-Chilcotin region, nearly 95% of the intermountain grassland has survived.

Climate

In a province known for lush coastal rainforests and towering mountains, the dry grasslands are often forgotten. However, these intermountain grasslands are part of the rich tapestry of ecosystems that is the heritage of British Columbia. Together with those in the Thompson, Nicola, Okanagan, Similkameen and Columbia valleys, the grasslands of the Cariboo-Chilcotin are born of the same high mountains and westerly air masses as are the rainforests. When moist air from the Pacific Ocean rises up the slopes of the Coast Mountains, it cools and drops more than 200 centimetres of precipitation

Bunchgrasses, such as bluebunch wheatgrass, distinguish the intermountain grasslands of British Columbia, Washington, and Oregon from the sod-forming grasslands of the Great Plains. Bunchgrasses are well adapted to the dry intermountain climate.

© Ordell Steen

Above: Fire sweeps over the Upper Grasslands much as it has for thousands of years, killing any small trees or shrubs that may have recently established in the grass.

Right: A fire scar at the base of an old Douglas-fir tree contains a record of past grassland fires. After each fire, the tree has attempted to grow over the enlarged scar, leaving an overlapping pattern.

When glacial meltwater drained into the Fraser and Chilcotin valleys, fine-textured deposits accumulated in the relatively still waters of glacial lakes, while gravels and stones accumulated on the bottoms of rushing rivers. These alternating deposits are evident on the slope across the valley.

Between fifteen and two million years ago, volcanic lava flowed over the Cariboo-Chilcotin landscape. When the lava cooled quickly, basalt rock columns were formed. Today they are exposed in many places in the grasslands.

Multiple valley terraces were formed when the river cut down through sediments deposited on the bottoms of ancient glacial lakes and rivers that filled the valley at the end of the glacial period. This erosion continues to this day.

per year onto the rainforests below. Wrung of its moisture by the cool temperatures, the air is dry as it slides down the east side of the mountains toward the Interior. As it descends, the air is compressed and warmed and its humidity is lowered. By the time the dry air reaches the Interior grasslands, it has the potential to pull more water from the soil than it returns as precipitation.

Only about 30 to 35 centimetres of precipitation falls here each year, less than in any other area of the Cariboo-Chilcotin. Air temperatures are the highest in the region. Forests cannot thrive here; only plants that can hold onto their moisture against the pull of the dry air, or that can avoid the drought by becoming dormant, survive in the grasslands.

Glacial erratics, dropped by the melting glacier about 10,000 years ago, are scattered over the grassland landscape. Unique communities of lichens, plants and animals take advantage of the microclimates they create.

Elegant orange lichens identify glacial erratics that are used by birds as perches and are enriched by their droppings.

Bunchgrass

Bunchgrasses are well adapted to survive in this dry climate. In the spring, snowmelt provides abundant moisture, and new green shoots grow quickly from the base of last year's bunch. During the summer drought, bunchgrasses become dormant. To conserve moisture, above-ground stems and leaves die back to the soil surface, yet remain erect, because of special rigid tissues in the leaves. The resulting funnel shape of the bunch directs rainwater to the plant's massive fibrous root system, while a small green growing point remains at or just below the soil surface, ready to produce new shoots when cool rains return in the fall or snowmelt once again provides moisture the next spring.

The Lower Grassland is the home of abundant big sagebrush, yet even here the plant community changes dramatically with changes in soil texture and slope aspect. Fine textured soils and north-facing slopes have the least big sagebrush and the most vigorous bunchgrasses.

THE LOWER GRASSLAND

Where fire has been absent, gnarled sagebrush shrubs form a grassland canopy like a dwarf forest. Big sagebrush is abundant in western North America, reaching its northernmost extent near Williams Lake.

As an adaptation to the dry climate, big sagebrush leaves are covered by fine hairs that reduce air movement and thus moisture loss from the leaf surfaces.

Fire in the Grass

Fire has been an important ally of the dry climate in maintaining the Cariboo-Chilcotin grasslands. According to the record contained in fire scars on trees, prior to the late 1800s fires burned the grasslands of the Cariboo-Chilcotin about every seven to fifteen years. Some of these fires were started by First Nations people to increase the abundance and growth of food and medicinal plants and to attract game animals to the new growth that followed a fire. Trees and juniper that invaded the grasslands were killed, while grasses and other plants that have their growing points at or below the ground grew back vigorously, fertilized by the ashes of the fire. After 1860, the frequency of fires decreased as cattle grazing removed the fuels necessary to carry fire, First Nations people were penalized for starting fires, and fire suppression reduced the size of wildfires. When the fires stopped, trees and shrubs invaded many cool, moist sites in the grasslands.

Shapes of the Land

In the nearly treeless grassland, the open landscape exposes the epic story of how the land was formed and sculpted. The expansive views offered by spreading vistas reveal the ancient shapes of the land. Bedrock outcrops stand boldly out of hills like exposed bones of the earth. Sometimes this bedrock is limestone, formed from calcareous sediments laid down in the ocean more than 200 million years ago when the western coastline of North America was in present-day Alberta. Most often, however, the exposed bedrock is of volcanic origin. Between 15 and 2 million years ago, volcanic lava erupted throughout the region. When it flowed across the land, it filled the valleys and covered most of the sedimentary bedrock. At the rim of many grassland valleys, the volcanic rock is exposed like the edge of a wave front, often in ordered, vertical columns.

The shapes of the current landscape were further sculpted during the great ice ages that began about two million years ago, when huge sheets of ice, up to two kilometres thick, spread over the land. The last ice sheet covered the Cariboo-Chilcotin plateau between twenty-five thousand and ten thousand years ago. Advancing and retreating, the ice sheets scoured the terrain and deposited

Above: Ladybug beetles are the principal predators of aphids.

Left: Formica ants feeding on the honeydew from aphids they are "farming" on a big sagebrush shrub.

Honeydew

Aphids are scale-like insects that feed on the sap of plants by inserting a stylet into the plant. They excrete excess carbohydrates as honeydew. Some species of ants, such as these formica ants, feed on the honeydew. In exchange, they protect these tiny insects from ladybugs and lacewings. Aphids have remained largely unchanged since they first appeared more than 200 million years ago.

A big sagebrush sends a taproot deep into the soil to find water.

Grassland Soils

The topsoil of the grasslands is fine sand and silt, called loess, that was deposited over the landscape by the wind just after the glaciers receded. Its dark colour is due to organic matter from the decay of fine grass roots. Enriched by organic matter, grassland topsoils are biological factories, humming with thousands of species and millions of organisms per square meter, including bacteria, fungi, protozoans, mites, and insects that break down organic matter, cycle nutrients and energy, control pests, and help plant roots take up nutrients and water. Beneath the topsoil lies a layer of accumulated white salts, left behind where percolating water has evaporated from its deepest penetration into the soil.

Crab spiders weave no webs, but are camouflaged to match their flower habitat. They may sit on the same spot for days, waiting for their prey to approach. This one sits on the head of a Hooker's thistle.

THE THISTLE WORLD

Thistles are prominent flowering plants that tower above most other herbaceous plants of the grasslands. There are two species of native thistles in the Cariboo-Chilcotin grasslands: Hooker's thistle (above) and wavy-leaved thistle. They occur mostly as scattered individuals and are habitat for many insects, such as blister beetles and crab spiders.

soil and rocks, leaving softly rounded hills and shallow basins. As the last ice sheet melted, large rocks and boulders, called glacial erratics, were left scattered in its wake. Shallow depressions, termed kettles, were formed where blocks of ice remained partially buried in the soil after this main sheet had melted away. They are now often wetlands, ponds, or the sites of aspen groves.

As the ice sheets melted, massive volumes of meltwater cut gullies down the slopes of valleys. Trapped behind ice dams, torrential rivers formed great lakes in the valley bottoms. Soil and gravel carried by the meltwater raised the bottom of the lakes. Sediments accumulated in deep layers, through which rushing rivers sliced as the ice dams melted and the lakes drained away. The history of the lakes and torrential rivers is told in the deposits that form the present terraces of the valley bottoms. Today, roads, pastures, and fields take advantage of the flat topography of these old lake bottom remnants.

After the glaciers melted, the new landscape had little vegetation to hold the soil. Winds lifted the glacial deposits and dropped

Top left: Blister beetles on wavy-leaved thistle. Adult blister beetles feed mostly on plant flowers or leaves and produce a toxic chemical that can poison livestock and blister human skin.

Top right: Grasshopper nymph on Hooker's thistle.

Bottom: The thistle flower head is a collection of hundreds of individual flowers.

them again to form a stoneless cap 30 to 60 centimetres thick over most of the grassland landscape. In some places, such as Farwell Canyon, wind-blown sands formed dunes that are still evident and active today.

The Three Grasslands

The three major ecosystems of the Cariboo-Chilcotin grasslands are commonly referred to as the Lower, Middle, and Upper Grasslands. Occurring in the bottoms of the Fraser and Chilcotin valleys at elevations below 650 metres, the Lower Grassland is the hottest and driest of the three. This is the land of abundant big sagebrush and prickly-pear cactus. Sagebrush scents the air with a sweet, pungent aroma, the very signature of a vast landscape that extends from the southwestern United States as far north as the Cariboo-Chilcotin. In the Lower Grassland, the bunchgrasses are widely spaced so that individual plants can draw precious moisture from a large area. Between the grasses and beneath the sagebrush, the soil is covered by a microbiotic soil crust of lichens and other organisms. Because the soil crust is very susceptible to trampling and other physical disturbances, it is substantially depleted in most areas.

In the Middle Grassland, few sagebrush shrubs are present and prickly-pear cactus is less abundant than in the Lower Grassland. A blanket of bunchgrasses — especially bluebunch wheatgrass, needle-and-thread grass, and junegrass — and abundant flowering herbs cover the land. In the summer, the bunchgrass blanket has a fuzzy look from the curled leaves of bluebunch wheatgrass. The microbiotic soil crust is present here, but not as extensively as in the Lower Grassland. The Middle Grassland is extensive in Churn Creek Protected Area and Junction Sheep Range Provincial Park.

The Upper Grassland is the ecosystem where the open grassland meets and merges with the vast Douglas-fir and lodgepole pine forests of even higher elevations. As winter snows are deeper and temperatures cooler than in the Middle Grassland, soils are moister in the spring and throughout the summer. In pristine Upper Grasslands, bunchgrasses completely cover the soil surface, completed by a colourful variety of flowering plants, such as sticky geranium,

Prickly-pear Cactus

Brittle prickly-pear cactus is well-adapted to the arid grassland climate. Water-filled stems, leaves reduced to spines, and pores that close during the day to reduce moisture loss, are all specialized adaptations for surviving the summer drought. Each June some plants open into extravagant yellow blossoms among the sagebrush of the Lower Grassland. Brittle prickly-pear cactus was an important food plant for First Nations people, especially in times of famine. Before the cacti were eaten, the spines were singed off with fire.

Upper left: A brittle prickly-pear cactus thrives on a parched slope in the Lower Grassland.

Left and above: The aggressive spines of prickly-pear cactus are reason to look carefully before sitting. The stem segments separate easily, often lodging in a boot or pant leg. When dropped again, segments easily root and establish new plants.

The best way we have found to experience the grasslands is at the most intimate level: by walking, observing, and listening.

Top: Vesper sparrows, a quintessential bird of the Cariboo-Chilcotin grasslands, build shallow nests of woven grass on the north or east side of the bases of large, old bunchgrasses, where the shade prevents the eggs from getting too hot.

Lower: The open mouths of young vesper sparrow nestlings form targets for adults bringing small insects as food.

Above: Ecologist Kristi Iverson examines a microbiotic soil crust in the Upper Grassland.

Right: In the Lower Grassland, the microbiotic soil crust is often best developed beneath big sagebrush plants, where it is protected from the hooves of grazing animals.

The Microbiotic Crust

The microbiotic soil crust is a complex biological community of lichens, cyanobacteria, mosses, fungi, and algae that covers and binds grassland soils, especially where grass cover is sparse, such as in the Lower Grasslands. It reduces soil erosion by binding soil particles together, helps to retain moisture in the soil by reducing evaporation, provides microsites for plant seeds to germinate and establish, and increases soil fertility by taking nitrogen out of the air and adding it to the soil in a form that plants can use. It includes a very large number of species, all of which are capable of drying out and suspending growth during droughts but responding quickly when moisture returns. The crust is very susceptible to mechanical disturbance and can be destroyed by human feet, cattle hooves, and off-road vehicles.

Various species of lichens (*colourful in the above image*) and rusty steppe moss grow on the mounded surface of a micro-biotic soil crust. The black crust is largely held together by cyanobacteria, which are single celled or filamentous bacteria. They are one of the oldest groups of organisms on earth.

cinquefoils, brown-eyed susan, and balsam-root. In contrast to the Middle and Lower Grasslands, short-awned porcupinegrass and spreading needlegrass cover gentle slopes and paint the grassland golden yellow in the fall; bluebunch wheatgrass is more abundant on warm slopes that face the afternoon or evening sun. The porcupine-grass-spreading needlegrass communities are unique to the Cariboo-Chilcotin. Further south in British Columbia and the U.S., rough fescue and Idaho fescue characterize

An expanse of Middle Grassland near the junction of the Fraser and Chilcotin rivers. At this elevation, sagebrush is virtually absent and the land is blanketed by a community of bluebunch wheatgrass, needle-and-thread grass, junegrass, pasture sage, and many broad-leaved flowering plants. Where gentle swales collect snow, porcupinegrass and spreading needlegrass flourish, creating bright yellow patches in the fall. Scattered Douglas-fir trees occur on rocky ridges and cool aspects.

THE MIDDLE GRASSLAND

The fruit of needle-and-thread grass, like other needlegrasses, is an adaptive and aesthetic wonder. Golden in colour, the fruit includes a 10 to 15 centimeter-long bent thread, or awn, and a sharp-pointed needle, containing the seed. Twisted tissues at the base of the thread swell and shrink at different rates as the humidity changes. With the thread braced against a stable object, the swelling and shrinking drives the needle into the soil (or into the sock of a hiker or skin of an animal). The seed is planted by the thread.

NEEDLE-AND-THREAD GRASS

A World of Butterflies

Any walk through the grasslands in spring or summer is accompanied by butterflies that rise out of the grass blanket, fly ahead, and land again. The warm sunshine and abundance of flowering plants makes this some of the best butterfly habitat anywhere. The best time to photograph butterflies is in the cool of the morning, when they perch with wings spread to absorb the sun's heat.

Above: A fritillary butterfly gathers nectar from alfalfa that has escaped to the grasslands.

Below: The Rocky Mountain apollo is a large, conspicuous butterfly that appears on the grasslands in late spring.

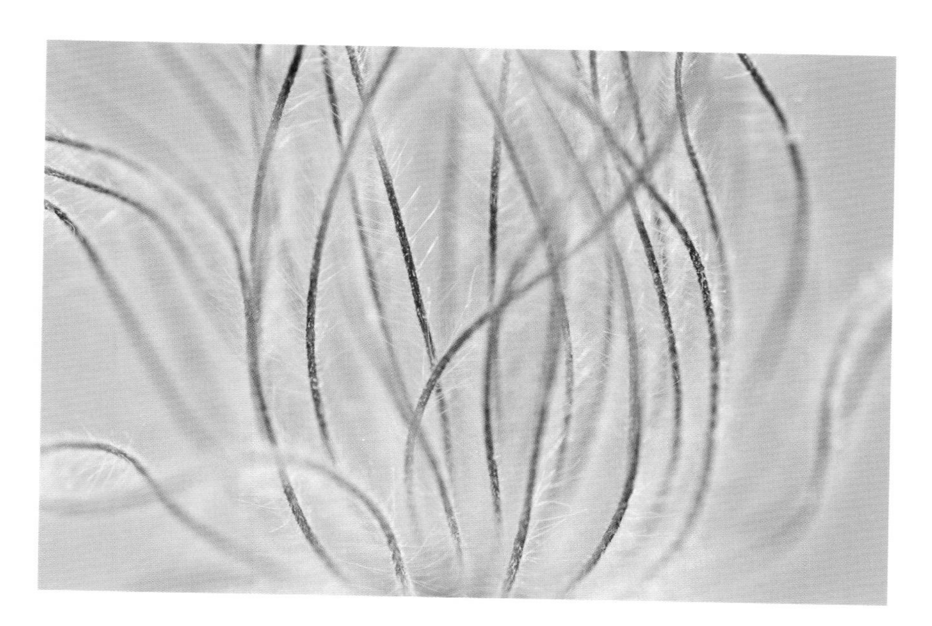

Above: The fruiting heads of old man's whiskers, sometimes called prairie smoke, wave in the breeze.
Below: Like feathers, the fruits are ready to be dispersed in the wind.

A hiker surveys a vigorous community of bluebunch wheatgrass blanketing an ancient, eroded slope in the Middle Grassland near Churn Creek.

pristine Upper Grasslands; these fescues are absent in the Cariboo-Chilcotin.

Seasons in the Grass

The grassland landscape in the Cariboo-Chilcotin changes dramatically with the seasons. In the winter, the land is quiet. Voles scurry through tunnels beneath the snow. Hunting short-eared owls may pounce on careless individuals that have emerged above the surface. California bighorn sheep paw through the thin snow cover for forage. Everywhere the blades of grass twist in the wind, drawing patterns on the thin snow cover. Snowdrifts form in shallow depressions and behind obstructions, storing moisture for spring.

Spring is marked by an explosion of life that revels in the warmth and moisture. Each day seems to bring an increased number of birds as they begin to establish territories and select nesting sites and mates. In March, the liquid songs of meadowlarks herald the awakening grasslands. They are soon joined by the low, burry songs of mountain bluebirds and a variety of ducks that jam themselves into patches of open water within partly frozen ponds. All will soon sort themselves out into mated pairs. On low ridges, sharp-tailed grouse return to traditional dancing grounds, called leks, to establish mating rights in the early morning light. Each year it seems there are fewer, even though they seem to be surviving better in the Cariboo-Chilcotin than they are further south. Spring is also marked by the appearance of rough-legged hawks migrating through the grassland on their way to arctic breeding areas. Other hawks that are present year-round, including red-tailed hawks and northern harriers, increase in number during the spring. The arrival of sandhill cranes, probably the world's oldest known surviving bird species, is heralded in April by penetrating guttural calls and aerial swarms of thousands of migrating birds. A few of these large birds nest in secluded wetlands within the grasslands.

Spring is a time when many plants flower and set seed quickly in order to take advantage of the moisture provided by the melting snows. One of the first flowers to appear in April is the sagebrush buttercup; it may even blossom just at the edge of melting snow banks. Soon after, the big

In the open grasslands and canyons, wildlife can be seen at great distances.

A herd of California bighorn sheep follows a trail across a steep slope used as escape terrain.

In the gently rolling plateau landscape of the Upper Grassland, open grassland merges with forests of Douglas-fir, lodgepole pine, and trembling aspen. These grasslands are relatively cool and moist, and grasses and other plants form a nearly complete blanket over the soil. At this elevation, bluebunch wheatgrass shares the slopes with porcupinegrass and spreading needlegrass. Small wetlands and aspen groves are common.

The Upper Grassland

Wetlands and their surrounding meadows and forests are critical areas for wildlife. They are used by more wildlife species, including more threatened and at-risk species, than any other grassland habitat.

yellow flowers of arrow-leaf balsamroot add brilliant splashes of color to the hill-sides. Balsamroot flowers may be so abundant that they nearly completely cover the ground, yet within a couple of weeks the yellow display is over and even the large, withered leaves are barely visible. Annual plants such as thread-leaved phacelia, quickly grow, flower and set seed before the intense heat of summer arrives. At upper elevations of the grasslands, the delicate lavender flowers of shooting star appear in April.

Summer is a less exuberant time in the grasslands, but a great time to discover the richness of life forms that have adapted to the dry climate. In late June, mariposa lilies create patches of mauve among the bunch-grasses, but then quickly die back to their underground bulbs to escape the summer drought. Other plants, such as big sagebrush and several species of pussytoes, reduce the effects of drought through leaves covered in hairs, which reduce air movement and resulting moisture loss. Big sagebrush also has both deep and shallow root systems. Deep roots draw moisture up at night into the shallower roots, from which moisture is taken up into the stem and leaves during the day. Stonecrop and prickly-pear cactus retain moisture by opening their pores only at night, when the humidity is higher. The carbon dioxide required for photosynthesis is then stored in an intermediate compound, before being released for use when the sun is shining and the pores are closed to prevent moisture loss.

Eared grebe, a common resident of grassland wetlands.

Fall brings bright colours to the grass-lands, especially the Upper Grassland, and often a regrowth and germination of bunch-grasses in the fall rains. The clashes of bighorn sheep horns can be heard in late fall when the rams battle to establish their social ranking.

Animals of the Grasslands

The most prominent mammals of the Cariboo-Chilcotin grasslands are California

Above: Aspen forests adjacent to wetlands are critical habitat for nesting waterfowl and other birds, including northern flickers.

Below: A University of Saskatchewan researcher takes measurements of a northern flicker (red-shafted race), as part of a study of their habitat requirements.

The Northern Flicker

Like other woodpeckers, the northern flicker hammers out nest cavities in trees but primarily feeds on ants on the ground, which it laps up with its long, barbed tongue. Although a few individuals winter in the Cariboo-Chilcotin, most arrive in March and leave again in September or October. Courtship rituals often involve one female and two males. The winning male sets up a nest in a tree cavity, often the same one it uses year after year, and males and females take turns sitting on the eggs. Straggler males are often able to woo females into laying another set of eggs in a second nest.

bighorn sheep, mule deer, and coyotes. California bighorn sheep is a subspecies of bighorn sheep that is slightly smaller than its Rocky Mountain counterpart and occurs mainly in grasslands. Junction Sheep Range Park and Churn Creek Protected Area are significant areas for the conservation of this subspecies. Sheep from the Junction area have been used to re-establish herds in six western states from which they had disappeared by 1900. California bighorn sheep lamb in rugged and remote areas of the Fraser and Chilcotin river valleys from late April to early June. Ewes with new lambs stay close to steep slopes used as escape terrain during the few weeks after lambing. As the lambs get older, groups of sheep move to less secure terrain with better forage. Poor lamb survival, due to diseases, lung parasites, and predation by cougars and coyotes, is a principal factor limiting bighorn sheep numbers. Mortality is increased by stress brought on by poor nutrition, trace element deficiencies, human harassment, and competition from livestock. California bighorn sheep are classified as at-risk in British Columbia.

Although rarely seen, badgers are increasing in the grasslands east of the Fraser River, and some have been observed west of the river. They typically dig their elliptical burrows in deep, stone-free, wind-blown soils. Due to loss of suitable habitat, hunting, and road-kill, badgers are endangered in British Columbia.

Small mammals in the native grasslands include montane voles, meadow voles, deer mice, common shrews, and many bats. The montane vole is common in the dry grasslands, especially where grass cover is sufficient to provide protection from predators such as red-tailed hawks. Voles are active year round, building tunnels under the snow in the winter and maintaining them through the summer by clipping off intruding grasses at ground level. They spend most of their lives above-ground, but dig burrows and make underground nests of dried grass. Although mated pairs do not share a nest, extended maternal families may be formed. Meadow voles are also common in the Cariboo-Chilcotin grasslands, especially on wetter sites.

In recent years, several species of bats

Meadowhawk dragonflies mate on a wetland sedge. Bright red when mature, these meadowhawks perch often and can easily be approached at the edges of ponds and wetlands.

Top: Red-winged blackbird.

Centre: Ruddy duck.

Bottom: Yellow-headed blackbird.

A quiet moment by a grassland pond or wetland brings entrance into the intimate world of birds, insects, and amphibians. Red-winged blackbirds belt out their familiar "kon-ka-reeee-a" songs from cattails. During courtship displays, male ruddy ducks slap the water rapidly with their blue bills. Near some shallow ponds, such as the one above, choruses of wood frogs fill the spring air with their deep duck-like calls.

The bright yellow of arrowleaf balsamroot flowers is a sure sign that spring has arrived, especially in the Middle and Upper Grasslands.

Balsamroot

Balsamroot was a primary food plant of First Nations people. Its young leaves were eaten raw or steamed, and the large taproots, which are sweet when cooked, were roasted or steamed, hung to dry completely, and then stored to be reconstituted later. The seeds were dried and pounded into flour. The young flower bud stalks were a favourite food after being soaked in water and peeled. Each large "flower" of balsamroot is actually a cluster of more than one hundred flowers. Those on the perimeter of the cluster have bright yellow, strap-shaped petals, adapted to look like petals of the entire cluster, while those in the center have much shorter, darker yellow petals.

The delicate, white blossoms of saskatoon add a spring-time splash to many ravines in the Lower and Middle Grasslands. Saskatoons are the most important native berry of First Nations people.

once thought only to occur further south have been discovered in the Cariboo-Chilcotin grasslands. These include the western small-footed myotis, the fringed myotis, and the spotted bat, which occur only in the grasslands, as well as the yuma myotis, which hunts along the rivers. Bats feed primarily on insects, with each species using a unique part of the grasslands. Bats are out mostly during the twilight and at night, spending days roosting under sloughing tree bark and in rock crevices.

Birds are nearly always visible in the grasslands. Birds that define the spirit of the grasslands include western meadowlark and vesper sparrow. Both nest on the ground, hidden within the bunchgrasses. Long-billed curlews also nest on the ground, preferably where the ground is level and the grass is short. Their loud cries are often heard before they are seen. Their long, curved beaks seem out-of-place in the grasslands. They are really shore-birds that winter on mud flats in the Gulf of Mexico.

Snakes of the Cariboo-Chilcotin grasslands include common garter snake, western terrestrial garter snake, rubber boa, gopher snake, and western yellow-bellied racer. Garter snakes are most common within 100 metres of ponds and wetlands, where they feed on toads, small mammals, and even small birds. The rubber boa is a true boa constrictor that is mostly nocturnal and seldom seen. Gopher snakes, which can be mistaken for rattlesnakes, have an aggressive behaviour and prey heavily on small mammals. The racer is an uncommon snake, found primarily in the Lower Grassland.

Western toad, Great Basin spadefoot, wood frog, Columbia spotted frog, and Pacific tree frog all occur in the Cariboo-Chilcotin grasslands. The spadefoot, a species considered vulnerable in British Columbia, is abundant in isolated locations here.

More than 50 species of butterflies occur in Cariboo-Chilcotin grasslands, where they thrive in the warm sun and on the rich variety of native flowering plants available for feeding. Most species depend on a particular host plant as food for their larvae. For example, blue coppers and square-spotted blues require parsnip-flowered buckwheat; painted ladies use the native grassland thistles; western whites and Stella

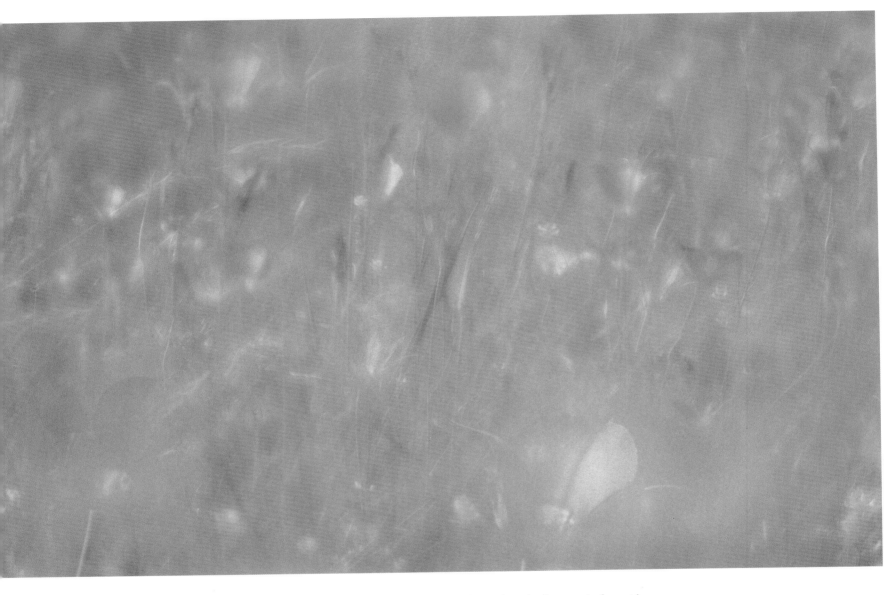

Mariposa lilies colour hillsides mauve in late spring, only to wither and nearly disappear before mid-summer. Mariposa lilies are less abundant than they once were, perhaps because of livestock grazing. Their underground fleshy bulbs were eaten both raw and cooked by First Nations people.

Shooting stars appear in clumps shortly after snowmelt, especially in the Upper Grassland, but also in some swales where snow lies late in the Middle Grassland.

The shiny, bright flowers of sagebrush buttercup are one of the earliest flowers to appear in the spring. All parts are mildly poisonous when fresh. Before hunting, First Nations people occasionally rubbed arrowheads with the sap of this plant.

Meadow salsify tracking the morning sun.

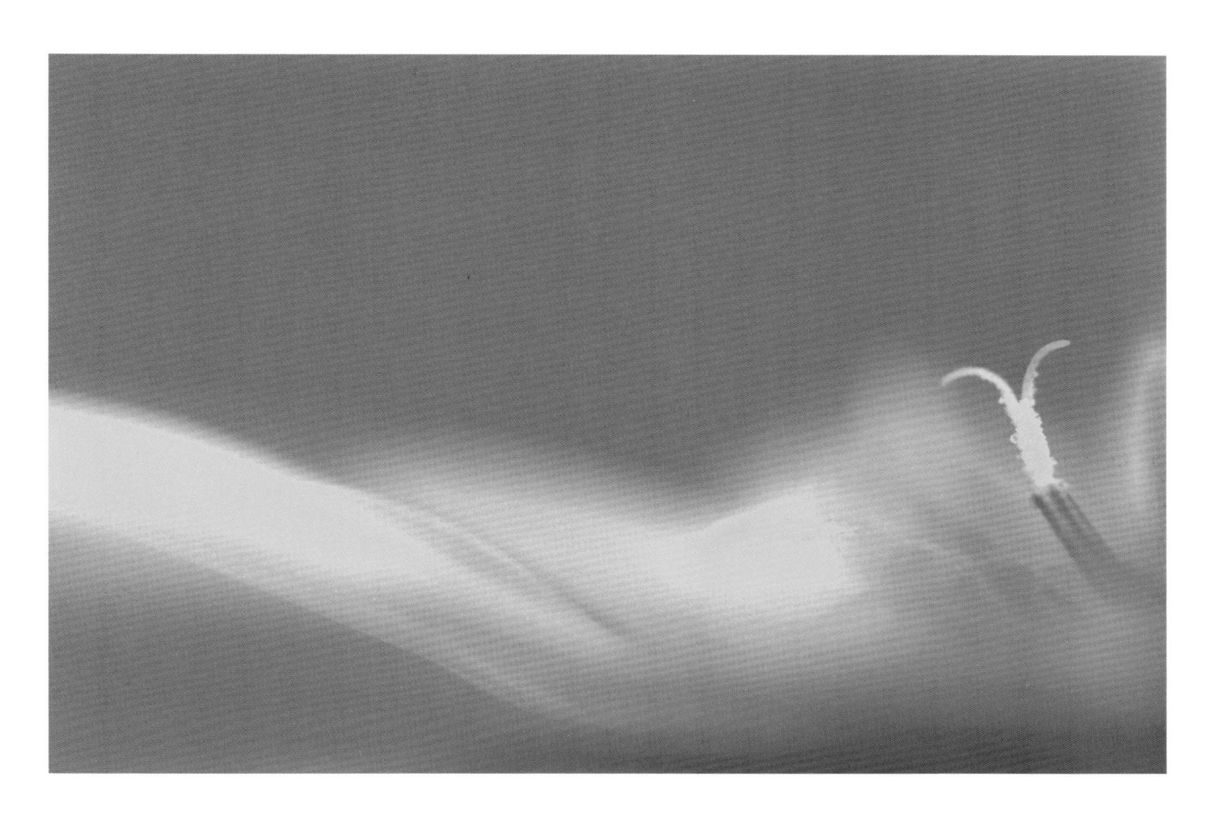

SALSIFY — FOLLOWER OF THE SUN

The yellow flowers of salsify (considered a weed by many) have received their name from the French salsifis, which is in turn abbreviated from the Latin solsequium, for sol (sun) and sequium (follower) — a plant that follows the sun. Its blossoms open rapidly in the morning sunlight, and then close up just as rapidly shortly after midday. Its seeds are perfectly adapted for dispersal by the wind and it quickly colonizes disturbed land. Salsify is not native to Canada.

The Cicada Story

A small hole about the diameter of a pencil eraser in the soil surface and a cast-off exoskeleton (outer shell) tell of the emergence of a cicada. Especially common in the sagebrush grasslands, cicadas live most of their time as nymphs under the ground, where they pierce the roots of plants and feed on their watery fluids. When cicadas are ready to emerge, they construct waiting cells just below the soil surface, where they await optimum conditions. As soon as they emerge, they cast off their exoskeletons to become colourful, flying, singing adults. Because they have good hearing, they are difficult to approach. After hatching, the nymphs fall to the ground and begin digging. Although some southern cicada species emerge simultaneously every thirteen or seventeen years, those in the Cariboo-Chilcotin grasslands emerge annually.

A puffball is the fruiting body of a spreading underground fungus.

orangetips rely on mustards; and the Rocky Mountain apollo depends on stonecrop. When grasslands are heavily grazed or cultivated and host plants are lost, many butterfly species disappear.

Ants are everywhere in the Cariboo-Chilcotin grasslands. They are a main predator of other insects, and the main movers of grassland soils, doing the work done by earthworms in wetter regions. The thief ant, which gets its name because it robs food from other ant species, occurs only in the grasslands. Although most species of ants spend time foraging on the surface, many live primarily underground, making nests under stones or bark and digging foraging corridors through the soil. Others, particularly formica ants, build thatched mounds on the surface. Streams of ants move on trails like rush-hour traffic into the city. Several ant species, including formica ants, field ants, and the tiny thief ant, tend

A pair of woodnymph butterflies perform a pas-de-deux around a wavy-leaved thistle.

The rubber boa is a true boa constrictor that feeds primarily on small mammals. Sometimes called the two-headed snake, it will often curl into a ball when threatened, hide its head, and expose its head-shaped tail.

aphids and other plant-sucking insects, both above- and below-ground, for the honeydew that they excrete. Some grassland ants raid the nests of other ant species to capture their eggs and larvae, which they carry back to their own nests, to be raised as slaves to perform work for the colony.

A rich variety of other insects and spiders are common in native Cariboo-Chilcotin grasslands, including blister beetles, ground beetles, damselflies, robber flies, cicadas, solitary bees, bee flies, and crab spiders.

Wetlands and Riparian Ecosystems

The greatest concentration of wildlife species within the grasslands occurs in its wetlands, ponds, streams, and adjacent riparian ecosystems. In the spring, wetlands and ponds are bustling with the activities and sounds of waterfowl courtship and nest-building rituals. In fact, the ponds and wetlands of Cariboo-Chilcotin grasslands support one of the greatest concentrations and most diverse assemblages of breeding waterfowl in British Columbia. The riparian shrub and deciduous tree communities adjacent to the wetlands and streams are critical habitat for a rich assemblage of plant and animal life. Many bird species such as the lazuli bunting, yellow-breasted chat, Lewis's woodpecker, and spotted towhee reach their northern distribution in grassland riparian ecosystems. Goldeneye and bufflehead ducks, northern flickers, mountain bluebirds and tree swallows nest in cavities in riparian trees.

Keeping the Spirit in the Grass

Unlike most grasslands of the world, those of the Cariboo-Chilcotin are little touched by urban and agricultural developments or the spread of invasive plants. Communities of bluebunch wheatgrass, big sagebrush, and California bighorn sheep, symbols of the once extensive intermountain grasslands, thrive here at their northern limits, even as they decline further south. In the soils, there are thousands of microbiotic life forms, including some of the oldest groups of organisms on earth. Both below and above ground, a profusion of insects display elaborate strategies for

Fairy Rings

Fairy rings are common on the Upper Grasslands. Each ring is created by a soil fungus that expands outward at a rate of perhaps five centimeters per year. Some large rings are probably more than 200 years old. As the fungus expands, it breaks down organic matter, adding a flush of nutrients that are taken up by grasses. Fruiting bodies of the fungus (facing page) appear at the edge of the fairy ring. The fungus seems to be beneficial to some grass species and detrimental to others. Cattle and even small mammals search out fairy rings for their very palatable forage.

survival. Many live among the porcupine-grass and spreading needlegrass, an ecosystem that occurs nowhere else. The ponds and wet-lands have one of the greatest concentrations of breeding waterfowl in British Columbia. Much of the rich tapestry of life in the grasslands remains unknown and discoveries of new species continue. They are a rich natural legacy and irreplaceable piece of our heritage that we can ill afford to lose.

— *Ordell Steen & Kristi Iverson*

Left: Kristi Iverson and Anna Roberts discover a slave ant species in a formica ant nest.

Right: The thatched mound of a formica ant colony is cleared of vegetation by ants and warmed by the sun.

Trails radiate outward from the thatched mounds of formica ants in the grasslands. Ants use the trails for quick travel to foraging areas and aphid farms, and for bringing food back to their mounds. When a foraging ant finds food, it drops chemicals, called pheromones, which other ants can follow to the food. A new trail may result.

Above: A montane vole races down a foraging path, back to the safety of its underground burrow.

Right: The world as a vole sees it. Without the protective cover of grass, voles would be visible to hawks, one of their main predators.

A red-tailed hawk hovers over the grasslands, hunting for small mammals. Red-tailed hawks are the most common hawk in North America.

A vesper sparrow brightens the grassland morning. Its song consists of two to four long clear notes followed by musical trills. Although common in the Cariboo-Chilcotin grasslands from April to September, populations throughout its range are declining as grasslands are converted to urban and agricultural use.

Above: Male horned larks sing like tinkling glass, from rocks, low mounds, or even while hovering. The brightly-coloured prairie race is most common here. The paler arctic race moves through in spring and fall. These birds often rise suddenly out of the grass or from the edge of a trail, only to quickly drop down again.

Right: A long-billed curlew watches for predators from its exposed rock perch. It is the largest shorebird in North America and is considered vulnerable to extinction in British Columbia. Long-billed curlews winter on the shores and marshes of the Gulf of Mexico but summer and breed in dry grasslands. Their loud "cur-lee" calls are heard for long distances.

Left: Naturalist Fred McMechan of the Scout Island Nature Centre in Williams Lake checks a mountain bluebird nest box on the grasslands as part of a regular monitoring program.

Right: Tree swallows will displace nesting bluebirds by building nests within cavities already occupied by bluebirds.

Mountain Bluebird Nest Boxes

Pairs of bird houses can be seen on many fence lines in the Cariboo-Chilcotin grasslands. The history of these nest boxes go back to a time when ranchers wanted to spread pesticides on the grasslands to control grasshoppers. To avoid pesticide use, naturalists put out nest boxes to increase the population of mountain bluebirds, a natural predator of grasshoppers. Nest boxes are placed in pairs because tree swallows will displace nesting bluebirds but will not live near another swallow family. Two closely placed boxes ensure that at least one will be available for bluebirds. Each spring, the nest boxes are cleaned to remove parasites. Bluebirds feed primarily by watching for insects from a perch or while hovering and then suddenly dropping on them.

A male mountain bluebird at its natural nesting site, a cavity in an old aspen tree.

All the varied habitats of the grasslands provide outstanding opportunities for discovering the natural world. The time taken here to observe, listen, and learn offers children life-long enriching rewards. Although each child's discovery is unique and personal, printed field guides and the experience of parents, naturalists, researchers, writers, and artists can enhance appreciation of the grassland.

LEARNING IN THE GRASSLANDS

Above: A male yellow-rumped warbler (Audubon race) adds a striking dash of yellow and black to an aspen grove.

Right: Red-naped sapsuckers build perfectly round holes as entrances to nesting cavities. To feed, they drill lines of small holes in aspen and willow trees and then return to feed on the leaking sap and the insects that are attracted to it.

An American avocet pair near the shore of a grassland pond. These elegant shorebirds feed in shallow water by sweeping their bills back and forth and stirring up aquatic invertebrates, which they capture by touch. The sexes are similar, but the female has a more up-curved bill. Considered endangered in British Columbia, American avocets are very uncommon in the Cariboo-Chilcotin, although the number of nesting pairs may be increasing.

A male avocet sits on three eggs, laid on the open ground without any nest structure. Chicks can leave the nest within 24 hours after hatching. After one day, they can escape predators by walking, swimming, or even diving.

The Secret Life of the Sharp-Tailed Grouse

Across western North America, the sharp-tailed grouse is an indicator species of grassland health. Largely due to habitat loss, the size and range of populations have declined dramatically over the last century. The Columbian sharp-tailed grouse, the sub-species found in the grasslands of the Cariboo-Chilcotin, was once widespread throughout the intermountain grasslands from southern California to central British Columbia. Populations have disappeared from most of the United States and southern British Columbia and have declined in Cariboo-Chilcotin grasslands. These amazing birds perform elaborate courtship dances, conducted on traditional dancing grounds called leks, to which they return annually. A lek is a nondescript area — usually on a ridgetop or a low knoll — distinguished only by a small area of exposed soil, a bit of stomped down grass, a few feathers, and bird droppings. The dances usually occur in the early morning and quickly break up when people approach. Females enter the lek to mate while the males are dancing, with the dominant male in the centre and the other males distributed around the edges. Genetic research indicates that the less dominant males are often just as successful at mating as the dominant ones.

Each male lazuli bunting develops a unique song by recombining song elements from older males in its neighbourhood. These birds occur primarily in riparian areas.

A killdeer sits on its eggs, nestled together on the bare ground near a small lake.
Killdeers usually draw predators away from their eggs by feigning a broken wing.

For small secretive birds, marsh wrens make a lot of sound. During the breeding season, they belt out their chattering trills all day and evening, like mechanical typewriters.. Living entirely within a marsh, they occasionally rise to the top of a cattail, but spend most of their time flitting through the tangle of stalks below.

A marsh wren gathers cattail seeds for a very cozy lining to its nest.

With great energy and persistence, marsh wrens build domed nests of sedges and grasses lashed to cattails or bulrushes above the water level. Males usually have several mates, each of which lines her own nest. In addition, males often build several dummy nests that may be used for roosting and to mislead or confuse predators.

A male Barrow's goldeneye, rejected by a female for another male, takes off to attack a male bufflehead. Barrow's goldeneyes are the dominant and most aggressive ducks of wetland ponds in the Riske Creek area.

Barrow's Goldeneye

The Chilcotin grasslands contain the largest breeding population of Barrow's goldeneyes in British Columbia. Goldeneyes build their nests in tree cavities. Females sometimes lay eggs in the nest of another goldeneye or even in the nest of another species. Once the ducklings emerge, they often leave their mother to join ducklings cared for by another female. After a female lays her eggs, the male flies north on a molt migration. Later the female follows the male to re-establish the pair bond. Together, they fly to southern tidal areas for the winter.

After mating, a male Barrow's goldeneye raises his head in a victory strut.

BURROWING OWLS AND BADGERS

Burrowing owls do not actually dig holes but rather rely on burrows created by other animals, especially badgers. This true grassland species once bred as far north as Williams Lake, but since 1928 only three natural nest sites have been found in British Columbia. All have since been abandoned. Non-captive owls are currently found only at re-introduction sites in the southern Okanagan and Nicola valleys. Reductions in the number of badgers, another at-risk grassland species in British Columbia, has been partially responsible for burrowing owl losses. Badgers have been reduced due to road kill, hunting, and loss of grassland habitat resulting from urban and agricultural development and forest encroachment. The number of badgers in the Cariboo-Chilcotin seems to be increasing. Whether this will mean a return of burrowing owls is unknown.

© Philippe Verkerk

Left: A northern saw-whet owl fledgling. One of North America's smallest owls, the saw-whet nests in cavities created by woodpeckers. These owls remain in the Cariboo-Chilcotin year round.

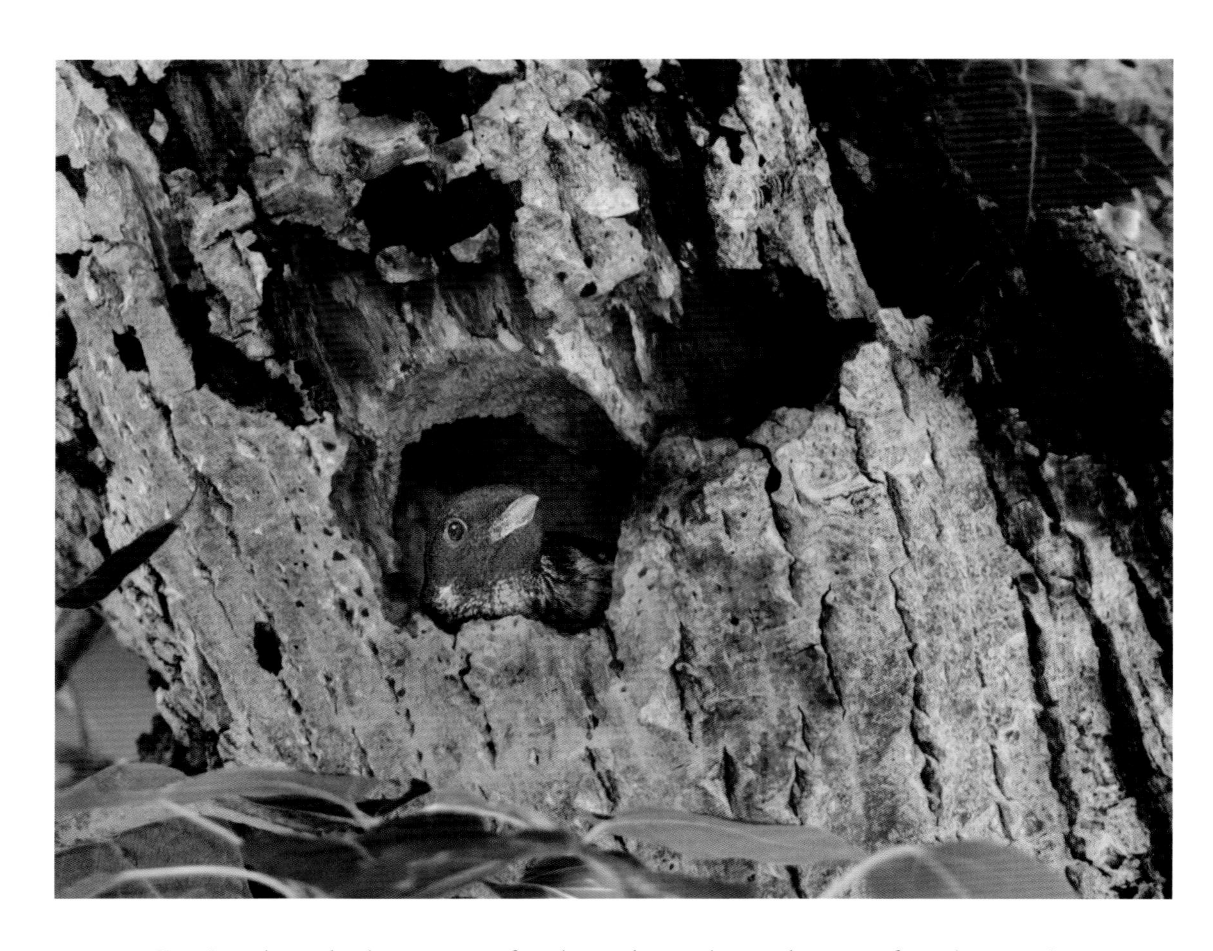

Since Lewis's woodpeckers are wary of predators, they won't enter their nests if people are nearby.

Lewis's Woodpecker

Where the terrace edge merged with the Fraser River, cottonwood leaves fluttered in the breeze and the sweet balsam scent of cottonwood sap permeated the hot air. We watched as a Lewis's woodpecker, unique among all woodpecker species, caught insects in the air over the willows and sagebrush. With a bug in its bill, the adult returned to feed the nestlings huddled in a cavity that was shaded and nearly hidden by the leaves of an old cottonwood tree. As the adult approached, gaping beaks emerged from the cavity. The adult woodpeckers, which usually mate for life, return each spring to these same remote cottonwoods. Because of the loss of big old trees required for their nests, their population is imperilled in British Columbia. Like many grassland birds in western North America, the population of Lewis's woodpeckers has probably declined by 60% since the 1960s.

After we spotted her nest, we were able to watch this female Lewis's woodpecker catching insects on the wing for her young.

© Anna Roberts

Bats in the Grasslands

Three of the twelve species of bats found in the Cariboo-Chilcotin region occur only in the grasslands. This western small-footed myotis is the smallest bat in British Columbia, weighing about five grams (slightly more than a nickel). This species hunts over sagebrush flats in the Lower Grassland and feeds on a variety of small insects, including moths, flies, and caddisflies. The fringed myotis is a less common grassland species that hunts along riparian areas. The spotted bat hunts along canyon walls and grassland valley slopes. Were it not for bats there would be many more night-flying insects.

Although the grasslands may seem like an unlikely habitat for bears, they can often be seen, especially in fall when they forage almost continually to build up fat for winter survival.

Relatively unchanged for many millions of years, sandhill cranes are one of the two oldest known surviving species of birds. Their rattling trumpet calls are often heard far overhead in the spring and fall as they migrate, often in flocks of thousands. Some of these shy birds stop to nest in the Cariboo-Chilcotin. Sandhill cranes mate for life and return each year to the same wetland to raise one or two young on a nest mound of bulrushes and cattails. Within hours of hatching, the young are able to forage for themselves. Within ten weeks, they are ready to fly south with their parents to maintain the ancient cycle of life.

THE ANCIENT ONES

Cultural History

The first people to inhabit the grassland were highly mobile, nomadic people who moved into the area about 10,000 to 9,000 years ago, shortly after the last ice sheet disappeared and the land was vegetated. Recent studies suggest that the first inhabitants were represented by a number of different cultures with various origins. These early people depended upon hunting deer, elk and other animals and the gathering of plants, together with limited fishing. By about 6,000 years ago, the nomadic pattern was beginning to change to a subsistence settlement pattern based increasingly on salmon fishing, together with deer hunting and plant gathering. Reliance on salmon fishing increased as the climate cooled and rivers became larger, resulting in expanded salmon spawning grounds. By about 3,000 years ago, the people had made the transition from highly mobile hunters to semipermanent village dwellers, settling during winter months in pit house villages along major rivers. During warmer months, they lived in temporary camps located throughout the grasslands and adjacent forest areas, where they hunted and gathered plants. During the late summer and fall, the people returned to the rivers to catch salmon, which was dried and stored, along with other food, in pits for the winter months. The pit house villages are the distinguishing feature of these first people, whose descendents are the Secwepemc.

During the thousands of years of Secwepemc and Tsilhqot'in culture on the grasslands, the world was a place of spirit, defined by stories. According to tradition, in the time before those stories, animals and people lived and spoke together, as brothers and sisters. Eventually it became the task of the trickster and transformer Coyote to transform this unified world into one divided between humans and animals, and in the pre-contact world of what would become British Columbia the people living in the grasslands were known

With paint mixed from fish oil and ground ochre, young First Nations people recorded their vision quests as pictographs on rock outcroppings.

Petroglyph along the Mid-Fraser River.

as the children of Coyote. The members of each nation considered themselves the descendents of a different one of Coyote's many wives.

Some of the most haunting stories of this long, human history of the grasslands concern people, often tricked by Coyote, who find themselves walking in an unfamiliar grassland, over which blows an endless, soft, warm wind. Soon they notice that the ground below them is full of holes, much like vole tunnels, and that the wind is pouring out of them. When they peer down through the holes, they discover that they are actually in the sky, and look down to see the Cariboo-Chilcotin grasslands far below. These stories draw heavily on ancient shamanic journeys to worlds of spirit. Hauntingly, for the people of the grasslands the other world was intimately tied to the one they knew everyday.

With the rush of gold miners to Williams Creek in 1858, the life of the grasslands was changed forever. By 1859, Kalalest, a Tsilhqot'in man, had been long-established with his family on the western bank of the Fraser River, just south of its muddy confluence with the turquoise, glacial waters of the Chilcotin River. There, Kalalest supported his family by raising and selling horses. Among the settlers who came with the Gold Rush were two former miners from Virginia via California, Thaddeus and Jerome Harper. Moving north from the Fraser Valley in search of new opportunities, they founded Harper's Camp (now the town of Horsefly) in 1859, and then roamed further west, to the Fraser River.

The Harper brothers approached the Fraser from the east, at Dog Creek, and rode north until they crossed to a Tsilhqot'in settlement. The Harpers fell in love with the land west of the Fraser, and persuaded the elderly Kalalest to sign much of it over to them. During their conversation with Kalalest, an imaginary line was drawn on the ground. After discussion, Kalalest agreed to let the Harpers use the land to the south of the imaginary line, while he was to keep the land to the north. Most of that land to the south of Kalalest's boundary eventually became the legendary Gang Ranch, the pride (and often the frustration) of many owners, including a Victorian newspaper magnate from London and a shiek from Saudi Arabia. It is currently run as an investment company.

During the first years of the Cariboo Gold Rush, the rest of the grasslands on both sides of the Fraser River were preempted into ranches as well, many of them settled by cowpunchers who had driven cattle up from Oregon and even San Diego. When they realized that the grasslands could easily support cattle and would make a better life than the hard life on the endless cattle drives, many decided to stay. Among these ranches, the Alkali Lake Ranch is the oldest ranch in Western Canada. The Empire Valley Ranch has been converted to a large protected area of intact grassland. The Spanish names in the vicinity of Williams Lake bear tribute to the Mexicans among the first ranchers.

Many native peoples of North America traditionally distributed wealth, formed social ties, and dissipated disagreements through annual gatherings of gambling and athleticism. For the Secwepemc, the annual gathering traditionally occurred in Williams Lake. By 1894 the last native gambling was made illegal in Canada. It was only with the formation of rodeos in the 1920s that grassland cultures found a new outlet for their cultural gatherings. Early histories of the Williams Lake Stampede, for instance, contain rich accounts of how entire First Nations families from much of the Cariboo and the Chilcotin made the journey by horse and wagon into Williams Lake, where they met, competed and gambled freely, and once again celebrated the movement and solidarity that remains at the heart of their ancient culture. Even in the 1970s, the Esketemc people spent the winter at their reserve village on a bench adjacent to the headquarters of the Alkali Lake Ranch, and their summers camping in "The Meadows," a network of savannahs in the hills behind. Now it is the horses of the Esketemc that run free on the grasslands, and the annual migrations of the Esketemc people are made out onto the grasslands to bring the horses and their spirit home, and then to return them again, in this culture that has never been broken.

Today, it is the fortunate confluence of the land sense of both ranchers and first peoples which have preserved the grass-lands and their spirit. Because of the large tracts of land necessary to support cattle in such a dry environment, much of the land has been saved from the fate of the Okanagan Valley, where similar large ranches were broken down a century ago into small orchard plots. Because of the continued honour that the Secwepemc and Tsilhqot'in people and the ranchers have maintained for the land, the spirit remains in the grass.

— *Harold Rhenisch*

The circular depressions in this image are the remnants of earth-covered log structures called pit houses. Pit house villages were primarily located near good fishing sites along the Fraser and Chilcotin Rivers and their tributaries. Some villages were likely occupied for thousands of years.

Herman Paul fishes for salmon at a traditional Secwepemc fishing hole.

Medicine Plants

Angela George, a Secwepemc elder of the Esketemc band, collects leaves of "red willow" (red-osier dogwood) with her great-great-granddaughter Teyah Dan. The Esketemc and other Interior Salish people collect many medicinal plants from the grasslands, including big sagebrush, rabbitbrush, wild tarragon, yarrow, and old-man's whiskers. Angela says that the "red willow," which grows along grassland streams and lakes, is good medicine; she frequently uses a decotion of leaves boiled in water as a hair and scalp treatment.

The horses of the Esketemc First Nations Band running free on the Lower Grasslands.

The Esketemc herd makes the long climb up from the Fraser River.

In the dreamtime, the trickster Sen'Klip left many children, whose various descendents became all the peoples of the Interior grasslands. Then he changed himself into a coyote. In that shape, he continues to roam the grasslands.

Home in the Grass

Coyote that passes through wheatgrass,
you are smoke that rises from the banks
of the San Jose River,
you are a year of sun in grass given spring light,
a patch of lichen
on an old stump fallen among grass stalks,
so that I have seen you and see
you now and do not see you
and have not seen you and see you
again. You are blue
bunched wheatgrass moving through needle
and thread grass, and the motion
of wind that is drawn through sedges
in a sweep of rain. So that I have seen you
and see you now, you are the first splash
of light, so that I see you, and have not
seen you, and there
in your earth, in story, you watch
for me and wait for me to come home
so that I see you again.

— *Harold Rhenisch*

The Empire Valley Ranch was purchased by the Province of British Columbia in 1998 and made part of the 36,000 hectare Churn Creek Protected Area, established in 1995. The objective of the entire protected area is to conserve and restore native grassland ecosystems while operating a viable cow-calf ranch. Ranch management is directed by BC Parks and the British Columbia Ministry of Forests and Range. The area includes Lower, Middle, and Upper Grasslands and has a long history of First Nations use.

A proud ranching family: John and Joyce Holmes and their three children, Hattie, Ellen, and Sam, of Empire Valley Ranch, Churn Creek Protected Area.

Haying at the Gang Ranch. Good stewardship means a balance of values, managed to be sustainable. It is a philosophy honed from a century and a half of isolation and independence.

Much of the history in the Cariboo-Chilcotin grasslands has taken place at the remote Gang Ranch.

Winter round-up, Riske Creek Ranch.

Branding at Riske Creek Ranch. In a land with few fences, cattle are still identified in the traditional manner.

An old fenceline runs lightly through open ranchland.

Grasslands are continually being consumed by housing developments like this one in Kamloops. Over 9,000 hectares of grasslands have been lost to urban and industrial development in the Thompson-Pavilion region alone. As population and agricultural pressures increase, losses on this scale could also occur in the Cariboo-Chilcotin.

Threats to the Grasslands

Our grasslands are under siege. Many of the grasslands of the Cariboo-Chilcotin have not yet succumbed, but threats felt strongly further south, such as urban development, invasive plants, abusive recreation, and intensive agriculture, are here and will continue to increase. Forest encroachment onto grasslands is already extensive.

The attractive open vistas of grasslands and their amenable climates have long attracted people. Now that populations are high and transportation cheaper, however, human presence and activity has become powerful enough to destroy, fragment, and degrade grasslands. Housing developments are the crest of an endless wave of people, all wanting to be the last to be able to build a home amidst such beauty. The Okanagan, the Thompson, and even the Cariboo's Williams Lake were once pristine grassland valleys. The erosion continues. In many places today, large grassland ranches have been divided into smaller acreages. Superficially, these ranchettes appear relatively natural, although in many cases they support horses, pets, and hobby farms, and are used so intensively that they simply become houses with ten-acre (or more) footprints.

In many grassland areas, alien plants — most often from western Asia — have displaced native plants and animals. First introduced to our grasslands through contaminated hay seed and remarkable in their hardiness and bounty, they produce vast quantities of seed that remains viable for decades. Lacking the natural enemies of their homelands, these invasive alien plants grow unchecked. Many, such as spotted knapweed, have attractive flowers; some, such as Dalmatian toadflax, have even been introduced through gardening; and new alien plants continue to arrive, including sulphur cinquefoil, a plant so competitive it can even displace other invasive alien plants, such as knapweed itself.

A particularly invasive annual grass is cheatgrass. Its low fuzzy single stems can be seen amongst native plants, or often in extensive stands of their own. This admirably adapted grass can make use of even the briefest periods of early-season moisture to germinate, grow, and set seed, completing its life cycle long before native grasses are ready to become dormant. In the process, it cheats our native plants of vital spring moisture.

Currently, the grasslands of the Cariboo-Chilcotin are relatively free of these noxious plants. Nonetheless, they are here, especially along our highways and roads, and they have the potential to spread at an exponential rate. We are the primary movers of these plants: on our vehicles, on domestic animals, on our pant legs, and sometimes by planting them unknowingly in our gardens.

These are not temporary scars. Because they attract repeat use, they last decades, if not indefinitely.

The use of human- and motor-powered vehicles for recreation continues to threaten the integrity of our grasslands. The paths and scars created by these vehicles readily attract further use, which deepens their impact. Because the tracks left by vehicular recreation gather run-off, they often cause severe erosion. In addition, weed seeds picked up in wheels and chains at these sites are often unknowingly planted at new locales the next day.

As human populations have grown, large ranches have remained the crucial barrier to the fragmentation and development of grasslands. As important as ranches have become to this vital ecosystem, however, if they are not properly managed they can also have negative consequences. Many ranchlands, for example, are overgrazed. On them, large bunchgrasses, such as bluebunch wheatgrass, are in decline, soils are

The impacts of increased population often extend far beyond urban boundaries. Dirt bikes and mountain bikes easily damage the fragile native plants and soils of the grasslands, replacing them with eroding soil and invasive plants.

THE CAPACITY TO RECOVER

Cattle have hammered this grassland down to bare dirt, pussytoes, pasture sage, and small grasses. Such degraded grasslands are good for neither cattle nor wildlife. Fortunately, as this grassland has not been invaded by alien plants it has the capacity to recover with proper management.

compacted, and the land itself is rendered vulnerable to erosion. In the early 1900s, many of our grasslands were grazed to nearly bare dirt. The positive message is that with the introduction of improved grazing practices many of these grasslands have recovered; others that have been more recently damaged can recover as well.

Once grasslands are tilled or developed there is much less hope of recovering them. South of the Cariboo-Chilcotin, many grassland areas have been converted to hayfields, ginseng plantations, orchards, and vineyards. In the Cariboo-Chilcotin, a cooler climate and more remote access has limited agricultural conversion. Current movements to introduce viticulture further and further north, however, constitute a new and pending threat.

Since the climate of the Upper Grassland is amenable to both trees and grass, it is highly susceptible to forest encroachment. In the past, fire swept away young trees, and thus maintained the grasslands and the adjacent savannah-like open forests. With the introduction of intensive grazing by domestic cattle in the mid- to late-1800s, the fine fuels that allowed these fires to spread largely disappeared. Soon after, First Nations peoples were no longer allowed to burn, and finally, in the 1950s, the province of British Columbia began actively suppressing wildfires. As a consequence, since the late 1800s successive waves of trees have encroached onto the grasslands and filled in the forests. Many former grassland areas are now thickly forested; in others, young seedlings — future forests — are growing amongst the grass. In fact, in some areas more than 30% of the once treeless Upper Grassland has disappeared in the last three decades alone.

Climate change will likely have significant effects on the grasslands of the Cariboo-Chilcotin. Although the area of grassland in the region is expected to increase over the long-term, the rate of climate change may be too rapid for native plant and animal communities to fully adapt to the new environment. As a result, weeds will likely become more abundant. The warming climate is also expected to stimulate human population growth and agricultural development, both of which will further threaten

This grassland has been lost to cheatgrass and tumble mustard.

these grasslands, at exactly the same time at which they will become increasingly important for conservation of grassland species moving northward from more southern areas.

Without our action, our grasslands will continue to be consumed and degraded by these forces. To save them, we must all change the way we locate and design housing developments, find means to keep working ranches working, map and treat invasive plants, and remove encroaching trees from this dwindling ecosystem. These are practical and reasonable steps, and they are within our means.

— *Kristi Iverson & Ordell Steen*

In healthy grasslands, robust bunchgrasses, grass litter, scattered cactus, and other flowering plants occur amidst a soil surface covered and bound together by mosses and lichens.

As these giant old trees die, there will be no other large trees to replace them.

The Lost Savannah

Prior to 1900, giant old Douglas-fir 'veterans' (above) were common on the edge of the grassland. Frequent fires kept out most of the thin-barked young trees, leaving the giants within a sea of grass and flowers. The savannah habitat supported many plant and animal species. Today surface fires no longer burn and the old veterans are choked by dense young trees that have shaded out grassland plants, insects, and birds. Grassland butterflies no longer find the host plants they once depended on and the species-rich savannah has been turned into a habitat for only a few species. Today, a fire in this forest would be a raging inferno rather than a cleansing fire moving through the grass.

Pines encroaching on porcupinegrass.

Waves of Trees

Historically, more grassland has been lost to forest encroachment in the Cariboo-Chilcotin than to any other threat. In the background of the image above, a dense forest of Douglas-fir trees occupies a site where about 50 years ago a long-established community of porcupinegrass and spreading needlegrass thrived. In the foreground, another wave of tree seedlings, this time lodgepole pine, have recently established on the grassland. Although grass still grows among the seedlings, soon there will be a forest here and our memory of this grassland will fade.

Fire or mechanical methods can remove these trees; repeated treatments will be needed as new waves of trees continue to establish.

The Complications of Fire

Land jurisdiction, smoke concerns, massive fuel loads in surrounding forests, dwellings, safety concerns, and different perspectives as to what values should be maintained in a landscape cause contemporary grassland burns to be both expensive and politically complex. This burn required a large staff, months of preparation, vehicles, water trucks, fire guard crews, and fire control contractors.

After a fire, the land turns into a garden, rich with the scents of new growth. Thread-leaved phacelia is a characteristic annual plant often found growing among the sagebrush. Its seeds, which survive for long periods in the soil, flourish after fire.

The Cariboo-Chilcotin grasslands (below) are relatively pristine. A hundred years ago, the South Okanagan (above) was a similarly undeveloped grassland. Without careful planning and protection, the biodiversity of the Cariboo-Chilcotin grasslands could be lost to similar development as that which has replaced almost all of the critical native valley-bottom grasslands in the Oliver-Osoyoos area.

In evening light I feel the spirit of the grass and of all the plants, birds, and animals that live there.

Reflections

I have been given a gift by those who walked the grasslands before me. Out there in silt canyons that few have walked, and on flats of big sagebrush in snow, fire, wind, and rain, and in all the long, almost endless days that make up the life of a nature photographer, up waiting for first light, and still with the lens to my eye as the last light fades from the sky, I find I have given my heart to the grass. It's out there now, with the first peoples, and with those who will come after me. By waiting patiently for the spirit of the grass to reveal itself to me, and remaining open for it to teach me, year after year I have grown more patient out in the grass. Now, at the end of this stage of my journey, I find that its spirit is within me, and within this book. I sought to capture it, but I never did. Instead, I found out how to give it freedom.

Along with my colleagues in this project, I hope that this book will serve as a source of inspiration for all your journeys into the grasslands, and the opening point for our most vital social discussions for centuries to come.

— *Chris Harris*

July 23, 2004. *As the sun set beyond this solitary tree in the Cariboo-Chilcotin grasslands, I discovered simple beauty. When I walked back to my camp for the night, I was at home on the grass.*

Further Information on British Columbia's Grasslands

Many resources are available for learning more about British Columbia's grassland ecosystems and species and for becoming involved in grassland conservation efforts. Only a few principal resources can be listed here.

The Grasslands Conservation Council of British Columbia (GCC) website (*www.bcgrasslands.org*) provides a comprehensive overview of British Columbia's grasslands, including grassland locations, species and ecosystems at risk, conservation values, and threats. It describes GCC programs to address grassland education, stewardship, and conservation issues, and provides links to several other sources for grassland information and conservation efforts. The BC Grasslands magazine, produced biannually by the GCC, is a forum for discussion of grassland ecology, management, and conservation.

Other websites with information related to grasslands and grassland conservation include those of the B.C. Ministry of Environment, Environmental Stewardship Division (*www.env.gov.bc.ca/wld*); The Land Conservancy (*www.conservancy.bc.ca*); The Nature Trust (*www.naturetrust.bc.ca*); the Nature Conservancy of Canada (*www.natureconservancy.ca*); the B.C. Chapter of the Canadian Parks and Wilderness Society (*www.cpawsbc.org*); the South Okanagan-Similkameen Conservation Program (*www.soscp.org*); the Invasive Plant Council of B.C. (*www.invasiveplantcouncilbc.ca*); and the Native Plant Society (*www.npsbc.org*).

Nature centres, especially the Scout Island Nature Centre in Williams Lake, the Allan Brooks Nature Centre in Vernon, and the Osoyoos Desert Centre in Osoyoos, are excellent resources for local information on grassland ecosystems and species. Staff in local offices of the B.C. Ministry of Environment and B.C. Ministry of Forests and Range, including those in Williams Lake and 100 Mile House, are also local grassland information resources.

The Grasslands of British Columbia is a technical report prepared (2004) for the GCC, that describes the location, climate, vegetation, soils, and wildlife of all major grasslands of the province. It is available for purchase through the GCC website.

British Columbia, a Natural History, Second Edition by R. Cannings and S. Cannings (Greystone Books, 2004) is a source of information on all major ecosystems of B.C., and includes a chapter on grassland ecosystems and species and the threats they face.

Plants of Southern Interior British Columbia edited by R. Parish, R. Coupé, and D. Lloyd (Lone Pine Publishing, 1996) is an easy-to-use field guide to plant identification, including grassland plants.

DISCOVER BRITISH COLUMBIA BOOKS™

Published by Country Light Publishing & Chris Harris Photography

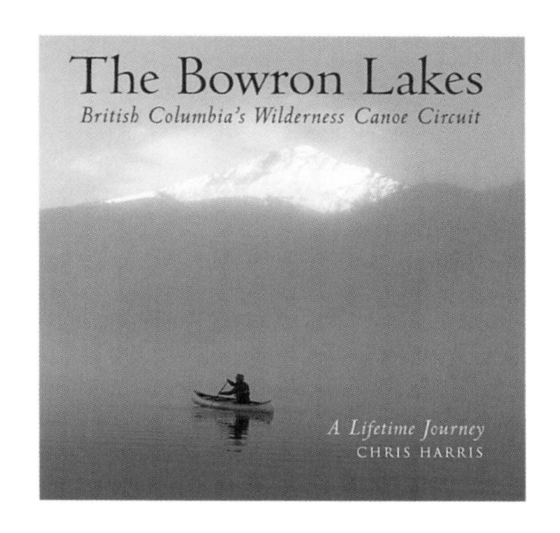

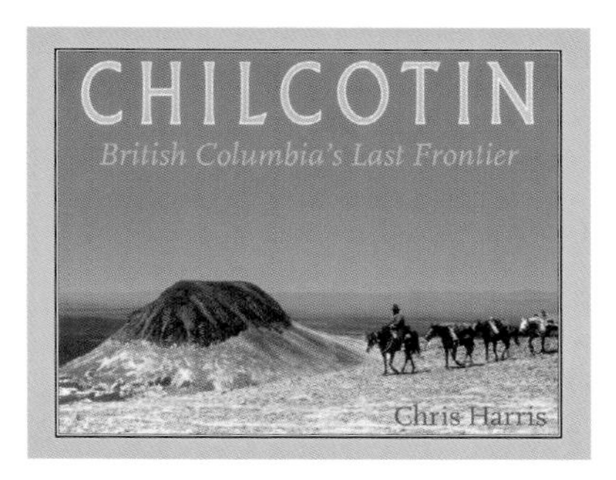

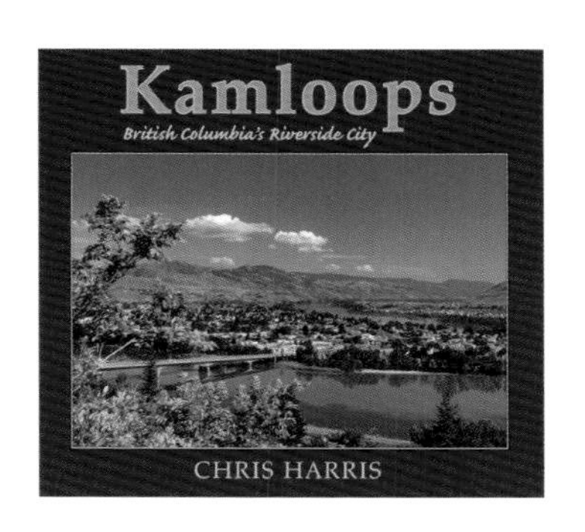

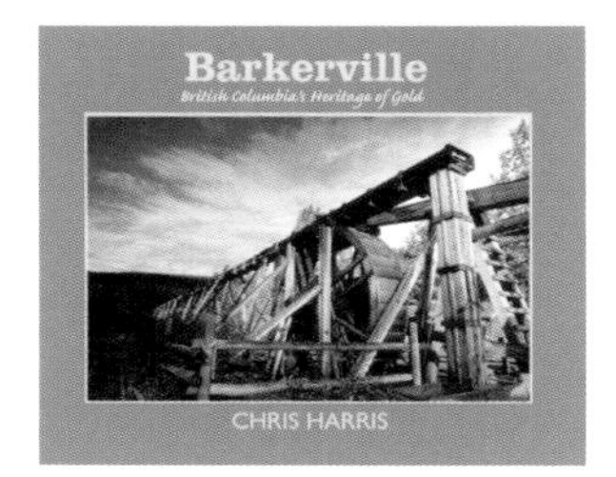

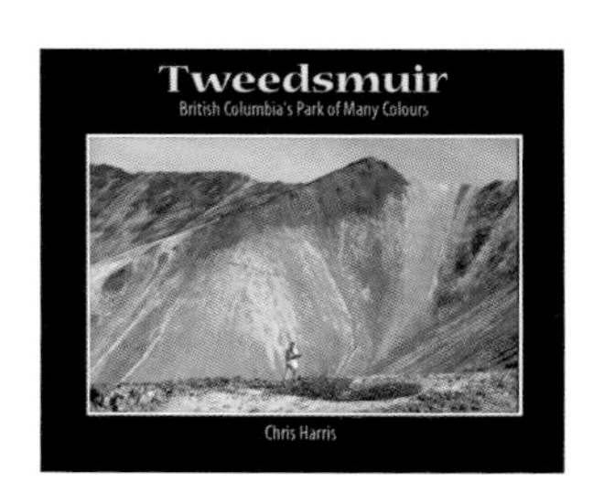

- Straw bale construction
- Fine art photographic prints
- Books, cards & gift items
- Audio-visual presentations
- Unique & world-class
- Art Collectors Welcome

Chris's fine art photography is exhibited in this unique and beautiful straw bale studio gallery. To view or purchase any Chris Harris books or prints, contact:

Country Light Publishing
Box 333, 108 Mile Ranch
British Columbia, Canada, V0K 2Z0

Ph: 250-791-6631 Fx: 250-791-6671

photography@chrisharris.com
www.chrisharris.com

Full moon over the grasslands.